Cambridge Elements

Elements in the Philosophy of Science
edited by
Jacob Stegenga
NTU Singapore

SCIENCE DENIAL

Post-Truth or Post-Trust?

Gabriele Contessa
Carleton University

Shaftesbury Road, Cambridge CB2 8EA, United Kingdom

One Liberty Plaza, 20th Floor, New York, NY 10006, USA

477 Williamstown Road, Port Melbourne, VIC 3207, Australia

314–321, 3rd Floor, Plot 3, Splendor Forum, Jasola District Centre, New Delhi – 110025, India

103 Penang Road, #05–06/07, Visioncrest Commercial, Singapore 238467

Cambridge University Press is part of Cambridge University Press & Assessment, a department of the University of Cambridge.

We share the University's mission to contribute to society through the pursuit of education, learning and research at the highest international levels of excellence.

www.cambridge.org
Information on this title: www.cambridge.org/9781009625272

DOI: 10.1017/9781009625326

© Gabriele Contessa 2025

This publication is in copyright. Subject to statutory exception and to the provisions of relevant collective licensing agreements, no reproduction of any part may take place without the written permission of Cambridge University Press & Assessment.

When citing this work, please include a reference to the DOI 10.1017/9781009625326

First published 2025

A catalogue record for this publication is available from the British Library

ISBN 978-1-009-62527-2 Hardback
ISBN 978-1-009-62529-6 Paperback
ISSN 2517-7273 (online)
ISSN 2517-7265 (print)

Cambridge University Press & Assessment has no responsibility for the persistence or accuracy of URLs for external or third-party internet websites referred to in this publication and does not guarantee that any content on such websites is, or will remain, accurate or appropriate.

For EU product safety concerns, contact us at Calle de José Abascal, 56, 1°, 28003 Madrid, Spain, or email eugpsr@cambridge.org

Science Denial

Post-Truth or Post-Trust?

Elements in the Philosophy of Science

DOI: 10.1017/9781009625326
First published online: October 2025

Gabriele Contessa
Carleton University

Author for correspondence: Gabriele Contessa, g.contessa@gmail.com

> **Abstract:** Over the past couple of decades, there has been growing concern about the alleged rise of various forms of science denial. But what exactly is science denial? How does it differ from ordinary scientific ignorance? Is it really on the rise? If so, what explain this trend? And what is so concerning about it in the first place? This Element has four goals. Its first (and least ambitious) goal is to is to bring some conceptual clarity by developing a clearer notion of science denial and gaining a better understanding of the concerns raised by the phenomenon it is meant to capture. Its second (and more ambitious) goal is to distinguish among several specific phenomena that are often conflated under the label "science denial." Its third (and even more ambitious) goal is to argue that none of these specific phenomena warrants all of the concerns that motivate its critics. Its fourth (and most ambitious) goal is to expose the inadequacy of a popular diagnosis of the epistemic malaise afflicting liberal democracies – *the post-truth diagnosis* – and to sketch an alternative to it – *the post-trust diagnosis*.

Keywords: science denial, science denialism, motivated reasoning, post-truth, post-trust, vaccine hesitancy, climate-change denial, creationism

© Gabriele Contessa 2025

ISBNs: 9781009625272 (HB), 9781009625296 (PB), 9781009625326 (OC)
ISSNs: 2517-7273 (online), 2517-7265 (print)

Contents

	Introduction	1
1	Why Is Science Denial Concerning?	3
2	What Is Science Denial?	12
3	What Explains Science Denial?	32
	Conclusion: Beyond 'Science Denial'	57
	References	63

Introduction

In a 2020 interview with *The Atlantic*, former US President Barack Obama sounded the alarm about an impending epistemological crisis:

> If we do not have the capacity to distinguish what's true from what's false, then by definition the marketplace of ideas doesn't work. And by definition our democracy doesn't work. We are entering into an epistemological crisis. I can have an argument with you about what to do about climate change. I can even accept somebody making an argument that ... it's too late to do anything serious about this .... I disagree with that, but I accept that it's a coherent argument. I don't know what to say if you simply say, 'This is a hoax that the liberals have cooked up, and the scientists are cooking the books. And that footage of glaciers dropping off the shelves of Antarctica and Greenland are all phony.' Where do I start trying to figure out where to do something?
>
> (Goldberg 2020)

While, presumably, most US citizens would not go to the extreme of suggesting that climate change is a hoax, recent estimates indicate that 42 per cent of them do not believe in anthropogenic climate change (i.e., the claim that climate change is primarily caused by human activity).[1] This stands in sharp contrast with the estimated 97 per cent of climate scientists who accept that claim.[2] While this phenomenon, which is often labelled 'climate-change denial', originated in the United States, similar discrepancies are now also observed in other countries.[3]

Climate-change denial seems to be one manifestation of a broader phenomenon, which, on a previous occasion, Obama had described as 'ignoring science':

> When President Kennedy set us on a course for the moon, I'm sure some made a serious case that it wouldn't be worth it. But I don't remember anyone *ignoring science*. I don't remember anyone saying the moon wasn't real, or that it was made of cheese.[4]

By 'ignoring science', Obama did not mean being ignorant of the relevant scientific claims. After all, we are all scientifically ignorant to some degree. Today, our collective scientific knowledge is so vast that none of us – not even the most knowledgeable scientist – can be expected to know everything currently known by science. For example, I do not know whether Higgs bosons can decay into photons, how many stars compose the Alpha Centauri star system, or what gene is associated with sickle cell anaemia, and, presumably, neither do you nor does Obama. Clearly, the kind of ordinary scientific ignorance in which

---

[1] (Marlon et al. 2023). [2] See, e.g., (Cook et al. 2013) and (Cook et al. 2016).
[3] (Fagan and Huang 2019). [4] As reported in (Holpuch 2014).

all of us partake is not the ignorance that concerns Obama, as it seems to be both ubiquitous and epistemically blameless. What seems to concern Obama is the kind of scientific ignorance of which climate-change denial is a species. It is the kind of scientific ignorance that is often referred to as 'science denial'.[5]

While Obama may be one of the most prominent critics of the phenomenon, he is by no means alone in thinking that science denial is concerning or that it is but one of the most serious symptoms of a deeper epistemic malaise afflicting liberal democracies. Over the past few decades, a growing number of academics, commentators, policymakers, and members of the general public have become increasingly concerned about the (alleged) rise of various forms of science denial, which, besides climate-change denial, are typically taken to include vaccine hesitancy, HIV/AIDS denial, creationism, flat-Eartherism, and, more recently, COVID scepticism.[6] These concerns have generated an extensive literature that includes academic articles,[7] opinion pieces,[8] as well as books with sensationalist titles such as *Denialism: How Irrational Thinking Hinders Scientific Progress, Harms the Planet, and Threatens Our Lives* (Specter 2009), *The Madhouse Effect: How Climate Change Denial Is Threatening our Planet, Destroying our Politics, and Driving Us Crazy* (Mann and Toles 2016), and *How to Talk to a Science Denier: Conversations with Flat Earthers, Climate Deniers, and Others Who Defy Reason* (McIntyre 2021). In what follows, I refer to the contributors to this literature (and to those who share their concerns) as '*counter-deniers*'.

But what exactly is science denial? How does it differ from ordinary scientific ignorance? And is it really as concerning as counter-deniers make it out to be? Unfortunately, counter-deniers have not given a clear and consistent set of answers to these questions. The first and least ambitious goal of this Element is to bring some conceptual clarity by trying to develop a clearer notion of science denial and gain a better understanding of the concerns raised by the phenomenon it is meant to capture. Its second (and more ambitious) goal is to distinguish several specific phenomena that are often conflated under the label 'science denial'. Its third (and even more ambitious) goal is to argue that none of these specific phenomena warrants all of the concerns that motivate counter-deniers. The fourth (and most ambitious) goal is to expose the inadequacy of a popular diagnosis of the epistemic malaise afflicting liberal democracies – *the post-truth diagnosis* – and to sketch an alternative to it – *the post-trust diagnosis*.

---

[5] As the by-line of the Guardian report states: 'President compares science denial to saying "moon is made of cheese"' (Holpuch 2014).

[6] The arguments in this Element focus exclusively on science denial and, *pace* (Bardon 2019, 49), they do not necessarily apply to other forms of denial, such as genocide denial (for a discussion of genocide denial, see (Altanian 2024)).

[7] See, e.g., (Hansson 2017), (Schmid and Betsch 2019), (Lewandowsky 2021), and (Jylhä et al. 2023).

[8] See, e.g., (Paul 2017), (Kwon 2019), (Friedman 2020), and (Mooney 2021).

The plan is as follows. Section 1 identifies two kinds of concerns that motivate counter-deniers – epistemic and practical concerns. Section 2 evaluates several 'thin' accounts of science denial – that is, accounts that describe science denial without trying to explain it – and finds them inadequate. Section 3 examines the most prominent 'thick' account of science denial, which attributes it to motivated reasoning.[9] While this account is more promising than the 'thin' accounts discussed in Section 2, it ultimately fails to vindicate the practical concerns of counter-deniers. Finally, the Conclusion argues that the term 'science denial' should be abandoned along with the oversimplified view of the relationship between science and society it presupposes.

1 Why Is Science Denial Concerning?

1.1 'Crazy' and 'Dangerous' Ideas

Counter-deniers typically take it for granted that science denial is a concerning phenomenon, and they usually make little or no effort to explain why it is so. However, before we can assess the validity of the counter-deniers' concerns about science denial, we must first understand them. In this section, I use the counter-deniers' scattered remarks on the topic to try to bring into sharper focus the exact nature of their concerns.

One of the clearest expressions of the counter-deniers' concerns can be found in the opening lines of philosophers Steven Nadler and Lawrence Shapiro's book *When Bad Thinking Happens to Good People* (which are also dramatically emblazoned in large orange letters on its red dustcover):

> Something is seriously wrong. An alarming number of citizens, in America and around the world, are embracing crazy, even dangerous ideas. (Nadler and Shapiro 2021, 1)

Like Obama, Nadler and Shapiro are not exclusively concerned with science denial – they are concerned with the broader epistemological crisis that is allegedly engulfing liberal democracies.[10] And they too, like Obama, consider science denial one of the most serious symptoms of that crisis.[11]

---

[9] The distinction between 'thick' and 'thin' accounts of science denial is possibly best explained by using a medical analogy. 'Thin' accounts conceive of science denial as analogous to a *syndrome* (i.e., a collection of symptoms that tend to cluster together but do not necessarily have a single underlying cause or set of causes), while 'thick' accounts conceive of it as analogous to a *disease* (i.e., a condition caused by a single underlying cause or set of causes).

[10] (Nadler and Shapiro 2021, 1).

[11] Of the seven examples of 'crazy, even dangerous ideas' listed by Nadler and Shapiro, three are beliefs typically associated with science denial – the belief that vaccinations cause autism, the belief that the scientific consensus on climate change is a hoax, and the belief that the 5G network

Nadler and Shapiro's opening remarks (along with many other scattered remarks by counter-deniers)[12] suggest that their concerns about science denial can be classified into two broad categories – epistemic concerns and practical concerns. Epistemic concerns include the concern that some people embrace 'crazy' ideas. Practical concerns include the concern that some of those 'crazy' ideas might be 'even dangerous'.[13] The rest of this section examines these concerns more closely. Section 1.2 focuses on the counter-deniers' practical concerns about science denial, Section 1.3 turns to their epistemic concerns about it, and Section 1.4 argues that the practical concerns should be given priority over the epistemic ones.

I should note that, while I have doubts about many of the factual and normative assumptions that underlie the concerns that motivate counter-deniers, the main goal of this section is not to assess the validity of those concerns but to better understand them, as it is important to develop a clearer understanding of the counter-deniers' views before evaluating or criticizing them.

1.2 Practical Concerns

The counter-deniers' practical concerns about science denial focus on the (alleged)[14] practical consequences of science denial. The broadest practical concern is that science denial might be (directly)[15] (ex ante)[16] practically

---

is responsible for the spread of COVID-19 (Nadler and Shapiro 2021, 1). Moreover, later in the book Nadler and Shapiro mention only two examples of ideas that are both 'crazy' and 'dangerous' – climate-change denial and vaccine scepticism (Nadler and Shapiro 2021, 18).

[12] See, e.g., (Bardon 2019, viii–xi), (Sinatra and Hofer 2021, 13), and (McIntyre 2021, 27–28).

[13] It is important to note that, as used here, the term 'counter-denier' applies only to people who share both sets of concerns outlined in this section. The use of labels such as 'science denial' is, thus, neither necessary nor sufficient to qualify as a counter-denier. Authors such as Neil Levy, who use those labels but don't share the counter-deniers' epistemic concerns (see, e.g., (Levy 2017)), do not qualify as counter-deniers, while authors such as Nadler and Shapiro, who rarely use those terms, qualify as counter-deniers because they share both the counter-deniers' epistemic concerns and their practical concerns.

[14] For the sake of brevity, I usually omit qualifications such as 'alleged' in what follows. However, the absence of these qualifications should not be understood as an endorsement of the counter-deniers' claims.

[15] A different worry is that science denial might be *indirectly* practically harmful. Even if a specific form of science denial is not itself directly practically harmful (it does not directly cause science deniers to make any decisions that have practically harmful consequences), embracing it might still cause science deniers to become more epistemically vulnerable to beliefs that are directly practically harmful (including those associated with other forms of science denial). In the terminology adopted here, if this is the case, then the first form of science denial would be (directly) epistemically harmful but not (directly) practically harmful – it would be only indirectly practically harmful. In the rest of this Element, 'practically harmful' means 'directly practically harmful' unless explicitly specified.

[16] Unless otherwise stated, throughout this Element, 'harm' refers to ex ante harms, where a decision harms someone ex ante if it increases their chances of incurring an ex post harm

harmful – that is, it might directly contribute to science denier's making decisions that have practically harmful consequences. McIntyre, for one, claims that science denial enables 'hundreds of thousands of people to refuse to vaccinate their kids, politicians to refuse to take action on climate change, and gun-toting protestors to parade during a pandemic' (McIntyre 2021, 28).[17]

It is helpful to distinguish three distinct levels at which science denial might be practically harmful – personal, interpersonal, and societal. Science denial is *personally* practically harmful to the extent that it directly contributes to science deniers' making decisions that have practically harmful consequences for them. For example, during the COVID-19 pandemic, many outspoken COVID sceptics died of the virus after refusing to take basic precautions against it.[18] While, in the absence of systematic empirical studies, we cannot be confident that their scepticism about COVID-19 played a causal role in their deaths, it is plausible to assume that people who underestimated the seriousness of COVID-19 were less likely to take precautions against it and more likely to contract the disease than people who took the disease seriously. An even clearer case of the practical harmfulness of science denial at the personal level might be the case of HIV/AIDS denial, as there is evidence that it results in many HIV-infected patients refusing antiretroviral therapies that significantly reduce mortality and morbidity due to HIV/AIDS.[19]

Science denial is practically harmful at the *interpersonal* and *societal* levels to the extent that it directly contributes to science deniers' making decisions that have practically harmful consequences for other people or, more generally, for the rest of society. For example, a parent's refusal to have their children inoculated with the MMR vaccine has harmful practical consequences for their children (as it increases their chances of contracting measles, mumps, or rubella), for those who come in contact with their children (especially those who are too young to be vaccinated or who cannot be vaccinated for medical reasons), and for society at large (as it contributes to undermining its herd immunity against those diseases).[20] Again, one of the clearest cases of interpersonally and socially harmful science denial might be HIV/AIDS denial, as there is evidence that HIV-infected HIV/AIDS deniers are more likely to have unprotected sex with HIV-uninfected partners, which is both

---

(even if they do not actually incur any ex post harm). Those who would rather think in terms of ex post harms should simply take 'harm' and 'harmful' to mean 'potential harm' and 'potentially harmful'.

[17] For similar practical concerns, beside the aforementioned remarks by Nadler and Shapiro, see, e.g., (Sinatra and Hofer 2021, 3).

[18] See, e.g., (Smith 2021a).

[19] See, e.g., (Bogart et al. 2010) and (Kalichman, Eaton, and Cherry 2010).

[20] See, e.g., (Porter and Goldfarb 2019) and (Feemster and Szipszky 2020).

practically harmful to their sexual partners and practically harmful for society, as it contributes to the spreading of HIV.[21]

Lastly, it is useful to single out a specific kind of practical concern at the societal level – namely, the concern that science denial that may have practically harmful consequences at the *political* level. The main concern here is that science denial can have detrimental effects on both political debate and policymaking.[22] For example, voters who do not believe in anthropogenic climate change are more likely to support political candidates who oppose policies aimed at mitigating anthropogenic climate change.[23]

Another political concern is that science denial can reduce voluntary compliance with beneficial recommendations or regulations. For example, vaccine-hesitant parents often circumvent vaccine mandates by claiming spurious religious exemptions.[24] Even if it is in principle possible for governments to coerce citizens into complying with their directives (by, e.g., issuing stricter childhood vaccine mandates), it seems to be neither desirable nor feasible for a liberal democracy to resort too extensively to coercive measures.

1.3 Epistemic Concerns

Let me now turn to counter-deniers' epistemic concerns about science denial. Epistemic concerns can be sorted into two broad categories – deontic and consequentialist. *Consequentialist epistemic concerns* are concerns about the epistemically harmful consequences of science denial.[25] Again, it is useful to distinguish three levels at which science denial can be epistemically harmful – personal, interpersonal, and societal. The main consequentialist epistemic concern at the personal level is that science denial can be epistemically harmful for the science deniers themselves, as those who embrace one form of science denial might become more *epistemically vulnerable* to other forms of science denial or, more generally, to unjustified beliefs.[26] The hypothesis is that, for example, becoming a flat-Earther might increase one's chances of becoming an anti-vaxxer, a climate-change denier, or a 9/11 truther. Several mechanisms may underpin the (alleged) increased epistemic vulnerability of science deniers. For example, it might be due to the interconnectedness of our beliefs, or to its effect on our epistemic character.

---

[21] See, e.g., (Desai and Majumder 2020).
[22] This seemed Obama's main concern in the first quote in the Introduction. For another expression of this form of the political concern, see, e.g., (Bardon 2019, viii–ix).
[23] See, e.g., (Tyson, Funk, and Kennedy 2023).   [24] See, e.g., (Williams et al. 2019).
[25] See, e.g., (McIntyre 2021, 27–28).
[26] See, e.g., (McIntyre 2021, 27–28). The phenomenon sometimes called 'conspirituality' (Beres, Remski, and Walker 2023) might be an example of this kind of epistemic vulnerability.

The main consequentialist epistemic concern at the *interpersonal* and *social* level is that science denial can cause an *epistemic contagion* – that is, the further spread of one form of science denial from one person to another and, more generally, through society. The hypothesis is that exposure science deniers' beliefs could convert others to science denial. The most obvious potential route of epistemic contagion is the direct route, which consists in one science denier persuading others to become science deniers.

Finally, here too, it is appropriate to distinguish one subspecies of epistemically harmful consequences at the social level – that is, those that are epistemically harmful at the political level. For one thing, science denial might contribute to undermining the kind of healthy public debate on which the proper functioning of liberal democracies depends by depriving the debaters of a shared ground of factual beliefs.[27] For another, science denial might affect education policy. For example, in the United States, creationists have engaged in various efforts to have creationism taught alongside evolutionary theory in biology classes.[28]

Deontic epistemic concerns about science denial focus on the epistemic conduct of science deniers. One possible deontic epistemic concern is that science deniers are *epistemically incompetent* – they do not have 'the capacity to distinguish what's true from what's false' (as Obama put it in his interview with *The Atlantic*). However, the most common deontic epistemic concern seems to be that science deniers are *epistemically irresponsible*.[29] This concern finds particularly clear and forceful expression in the following passage from Nadler and Shapiro:

> A simple, if somewhat brutal, diagnosis of the current state of affairs in America is this: a significant proportion of the population are not thinking reasonably and responsibly. The real problem is not that they lack knowledge, education, skill, or savvy. Acting on incomplete knowledge or without the requisite skills can doubtless lead to disagreeable consequences. However, a person who does so might be blameless – morally blameless, if she really could not have done otherwise, and even epistemically blameless, if she could not possibly have known better. ... On the other hand, bad thinking, as we understand it, is a character flaw deserving of blame. Unlike ignorance or lack of intelligence [bad thinking] is generally avoidable. People who think badly

---

[27] This seems one of Obama's concerns in the interview quoted in the Introduction. For a similar concern, see, e.g., (Bardon 2019, viii–ix).
[28] See, e.g., (McIntyre 2021, 27–28).
[29] In this Element, I assume that epistemic agents are epistemically responsible to the extent that they fulfil their epistemic responsibilities and/or obligations, and they comply with any relevant epistemic norms.

do not have to think badly. They may be – or, at least, should be – perfectly aware that they are forming and holding beliefs irrationally and irresponsibly, and even doing so willfully. (Nadler and Shapiro 2021, 2–3)[30]

The concern seems to be that phenomena such as science denial result from an epistemically blameworthy kind of epistemic irresponsibility rooted in some kind of epistemic vice. While Nadler and Shapiro call the alleged epistemic vice 'epistemic stubbornness', others seem to think of it as something closer to what Quassim Cassam (2019) calls 'epistemic insouciance', which is an indifference to (or lack of concern for) the truth.

These deontic epistemic concerns converge with *the post-truth diagnosis*, which attributes the epistemic malaise afflicting liberal democracies to a lack of interest in (or concern for) the truth (or any other epistemic goals) on the part of a growing number of their citizens.[31] Since science denial is often considered one of the key symptoms of that malaise, science denial can be regarded as a crucial test case for a post-truth diagnosis. If it turns out that, as many counter-deniers seem to believe, science denial is best explained in terms of science deniers' lack of interest in (or concern for) the truth, then this would support a post-truth diagnosis of the epistemic malaise afflicting liberal democracies.

While it is too early to try to assess the validity of the counter-deniers' concerns, a few remarks about the significance and the legitimacy of deontic epistemic concerns are in order. The first remark is that deontic epistemic concerns are not concerns about science denial per se – they are concerns about the deeper epistemological problems that (supposedly) underlie phenomena such science denial (such as our fellow citizens' (alleged) epistemic incompetence or their (alleged) lack of interest in (or concern for) the truth). If so, then science denial is merely a symptom of deeper problems.

The second remark concerns the political legitimacy of questioning epistemic competence in liberal democracies. Historically, accusations of epistemic incompetence have served to marginalize and disenfranchise already disadvantaged social groups (including women, ethnic and racial minorities, people with disabilities, and the poor).[32] Even today, claims about the alleged epistemic shortcomings of our fellow citizens are often used to undermine the ideals of democracy and challenge the political authority of democratic governments.[33] For these reasons, in a liberal democracy, citizens must operate on the

---

[30] For similar concerns, see also (McIntyre 2019, 151) and (Bardon 2019, 102).
[31] See, e.g., (Ball 2018), (D'Ancona 2018), and (McIntyre 2018). For critical discussions, see (Habgood-Coote 2019) and (Hannon 2023). It is significant that one of the most vocal counter-deniers, Lee McIntyre, is also a staunch supporter of the post-truth diagnosis and that most supporters of the post-truth diagnosis, including McIntyre, consider science denial as one of the main symptoms of post-truth.
[32] See, e.g., (Gould 2006). [33] See, e.g., (Brennan 2017).

presumption that our fellow citizens are epistemically competent unless proven otherwise.[34] Moreover, as liberal democracies become more polarized, we must exercise particular caution in levelling accusations (or even insinuations) of epistemic incompetence or poor epistemic character, as such charges can all too easily be used to dismiss without due considerations the sincerely held beliefs and genuine concerns of those who disagree with us. Of course, this does not mean that we should naïvely assume that all our fellow citizens invariably act in good faith. For instance, it is undeniable that some actors deliberately spread disinformation on scientific topics to advance their own social, economic, or political agendas.[35] However, there is a crucial distinction between challenging the motives of those who act *in bad faith* and impugning the epistemic competence or the epistemic character of those who act *in good faith*.

The third remark is that it is unclear whether deontic epistemic concerns are compatible with one of the pillars of liberal democracy – the principle of epistemic tolerance, which holds that everyone is entitled to their own beliefs, no matter how wrong or bizarre they may seem to others. To an atheist, the religious beliefs of Christians, Muslims, or Hindus might seem no less bizarre or unfounded than the beliefs of flat-Earthers seem to counter-deniers. However, in a liberal democracy, everyone is entitled to their own beliefs, no matter how bizarre or unfounded those beliefs might seem to some of their fellow citizens, and this seems to apply as much to the sincerely held beliefs of flat-Earthers as to those of Mormons or Jehovah's Witnesses. Of course, this does not mean that citizens of a liberal democracy must refrain from trying to rationally persuade other citizens to change their mind. However, rational persuasion differs significantly from publicly impugning someone's epistemic competence or responsibility.[36]

Counter-deniers may reply that political rights come with a host of associated responsibilities and that, in the case of freedom of opinion, the relevant responsibilities are primarily epistemic responsibilities.[37] On this view, those who fail to fulfil their epistemic responsibilities essentially forfeit their right to their own opinion. One problem with this view, however, is that it seems to uphold epistemic tolerance in theory while undermining it in practice. What would prevent an atheist (or theist) from arguing those who disagree with them forfeit their right to their own religious beliefs because they fail to fulfil their epistemic responsibilities? Moreover, even if freedom of opinion were conditional on

---

[34] I take this to be reflected in the fact that liberal democracies typically set extremely high legal standards for declaring someone legally incompetent (where epistemic competence is a necessary component of legal competence).

[35] See, e.g., (Michaels 2008) and (Oreskes and Conway 2010).

[36] I take it that one of the problems with so-called New Atheists, for example, is that they often seem to cross that line.

[37] See, e.g., (Norman 2021).

epistemic responsibility, it would still be incumbent on counter-deniers to substantiate their claim that science deniers do, in fact, fail to fulfil their epistemic responsibilities. As I argue in this Element, that task is much harder than counter-deniers typically seem to think.

1.4 On the Primacy of Practical Concerns

While this brief survey is unlikely to be exhaustive, it provides a preliminary taxonomy of the concerns that seem to animate counter-deniers, which, in turn, might help bring into sharper focus the nature of the phenomenon that primarily concerns them. While I am not persuaded of the validity or legitimacy of most of these concerns, I want to set my worries aside for the moment. Instead, I want to argue that, even if all counter-deniers' concerns were valid and legitimate, they would not be all equally important or urgent – the practical concerns would have priority over the epistemic ones. In fact, I would argue that, ultimately, it is appropriate to be concerned about science denial only to the extent that it is practically harmful.

Let me start with what I have called deontic epistemic concerns. Deontic epistemic concerns include the concern that science denial might be a symptom of a deeper epistemic ill, such as the (alleged) epistemic incompetence of science deniers or their lack of concern for the truth. Those who doubt the primacy of the practical concerns might note that deontic epistemic concerns have nothing to do with the practical consequences of science denial, as they are explicitly purely epistemic concerns. On closer scrutiny, however, this claim does not seem correct.

To see why, suppose, for example, that all forms of science denial were completely practically harmless.[38] Would deontic epistemic concerns still be legitimate? Counter-deniers might insist that, even if all forms of science denial were completely practically harmless, science deniers would still be epistemically blameworthy for their epistemically unjustified beliefs. They might insist that, when science deniers believe that the Earth is flat or that the MMR vaccine causes autism despite all of the evidence to the contrary, they are violating basic epistemic norms, and the violation of those norms would still be epistemically bad even if the resultant beliefs were entirely practically harmless.

---

[38] Science denial would be entirely practically harmless if it was neither directly nor indirectly practically harmful, where a form of science denial is directly practically harmful insofar as it directly causes science deniers to make practically harmful decisions and it is indirectly practically harmful insofar as it causes science deniers to become more epistemically vulnerable to other forms of science denial (or any other epistemically unjustified beliefs) that are directly practically harmful.

Personally, I do not find this line of reasoning persuasive, as I doubt that epistemic norms have any independent normative force.[39] However, since I cannot defend such a general epistemological view here, I will not pursue this objection. Instead, I will argue that, even if epistemic norms do have independent normative force, it does not follow that their violation is an appropriate object of concern on the part of the counter-deniers. Let me briefly mention two reasons why this is so.

First, just because my fellow citizens do something that I believe to be intrinsically bad, it does not follow that it would be appropriate for me to be concerned about it. Suppose, for example, that I believe that all lies are intrinsically morally bad. Does it follow that I should be concerned that many people lie? Clearly, it does not. After all, despite my beliefs about the morality of lying, I concede that many lies are perfectly harmless and that, in fact, some lies might even be beneficial (e.g., an innocuous lie told not to hurt a friend). The fact that something is intrinsically morally bad does not mean that it is all-things-considered bad, let alone an appropriate object of concern for others.

Second, all of us have some epistemically unjustified beliefs and most of us act (epistemically) irresponsibly from time to time. However, as long as our unjustified beliefs are practically harmless, our epistemic irresponsibility does not seem to be an apt object of concern on the part of others. For example, most people I know (myself included) seem to have relatively strong beliefs about the personality, the personal lives, or the moral character of various celebrities. These beliefs are usually based exclusively on what we have learned about those celebrities through the media. Most likely, most of those beliefs are epistemically unjustified (as they are based on little low-quality evidence) and, in holding them, we are likely acting in an (epistemically) irresponsible or blameworthy manner (since our beliefs should fit our evidence). However, no one seems particularly concerned about our epistemic failings in this regard, and rightly so. Even if most of our beliefs about celebrities are epistemically unjustified, our epistemic shortcomings do not seem grounds for concern if those beliefs are practically harmless.[40] Similar considerations apply to more outlandish beliefs, such as the belief that people's astrological signs affect their personalities.

---

[39] I am sympathetic to epistemic instrumentalism – the view that epistemic rationality is just a form of instrumental rationality. I believe that it is rational for us to comply with epistemic norms only insofar as doing so facilitates the achievement of our epistemic or practical goals. For an overview of the debate (as well as for two contrasting views on epistemic instrumentalism), see (Carter 2024) and (Worsnip 2024).

[40] This is not to deny that our society's obsession with celebrities might be harmful in many other ways. For example, it might harm the celebrities themselves. However, the point is that, if our beliefs themselves are practically harmless, the fact that we have those beliefs should not be anyone's concern.

These beliefs seem no more epistemically justified than those of flat-Earthers, and those who hold them do not seem more epistemically responsible than flat-Earthers. However, no one seems to be particularly concerned about those beliefs, and rightly so (as long as the relevant beliefs are practically harmless). Why would the same considerations not apply to the beliefs of flat-Earthers?

As far as I can see, the most plausible answer to this question is that the beliefs of flat-Earthers and other science deniers, unlike beliefs about celebrities or beliefs about astrology, are *not* completely practically harmless – they are either directly or indirectly practically harmful. On closer inspection, thus, deontic epistemic concerns seem to be largely a distraction – the most serious concern about science denial is not that science deniers are epistemically incompetent or irresponsible but that their beliefs are directly or indirectly practically harmful. While it is still too early to assess the validity of the practical concerns, our preliminary survey suggests that it is plausible that *some* forms of science denial, such as vaccine hesitancy or HIV/AIDS denial, may be directly practically harmful and that it is not entirely implausible to suspect that forms of science denial that do not seem directly practically harmful (such as flat-Eartherism) may nonetheless be *indirectly* practically harmful, as they make those who embrace them more epistemically vulnerable to forms of science denial (or other unjustified beliefs) that *are* directly practically harmful.

2 What Is Science Denial?

2.1 Do We Need an Account of Science Denial?

This section explores one of the most basic questions about science denial – that is, what is science denial? Unfortunately, counter-deniers do not give a clear or univocal answer to this question. In fact, most counter-deniers do not even try to answer it at all. Instead, they simply rely on an open-ended list of prototypical forms of (alleged) science denial, which typically includes climate-change denial, vaccine hesitancy, creationism, or flat-Eartherism. But are such open-ended lists not sufficient? Why would we need an account of science denial? After all, we successfully use many ordinary terms (such as 'chair', 'cat', or 'game') without ever being given any explicit definition of them. We have all learned what a chair is by being exposed to a variety of chairs and being told that they were chairs. Why would the same not apply to 'science denial'? Moreover, don't we all know science denial when we see it?

There are several reasons to find this approach unsatisfactory. Let me mention four. First, it is not clear what is included in the open-ended list and what is not. While we typically agree on whether a given object should be classified as a chair, counter-deniers do not even agree among themselves on who should be classified as a science denier. Journalist Michael Specter, for one, interprets the concept so broadly as to include among science deniers not only anti-vaccination conspiracy theorists but also people who eat organic food or take multivitamins.[41] Presumably, most people (including most counter-deniers) would disagree with Specter on this point.[42] Whatever one may think of the health benefits of consuming organic foods or taking nutritional supplements, it seems implausible to tar the people who eat organic food or take multivitamins with the same brush as the most radical anti-vaxxers. However, without a clear notion of science denial, there is no principled way to determine where exactly to draw the line.

Second, without a clear account of science denial, it is unclear how prevalent science denial is or how concerned we should be about it. If, for example, we were to follow Specter in classifying anyone who eats organic food or takes nutritional supplements as a science denier, then science denial would be a very widespread phenomenon indeed. However, it would also not seem a particularly concerning one. After all, multivitamins and organic carrots might not have any beneficial effects on the well-being of those who consume them, but they do not seem directly or indirectly practically harmful either. The belief that organic carrots are healthier or more environmentally friendly than non-organic one appears to be practically and epistemically harmless even if false.

Third, even if counter-deniers could offer an exhaustive list of forms of science denial, this would only shift the problem – the problem would then be how to correctly use labels such as 'climate-change denier' or 'anti-vaxxer'. For example, what qualifies someone as a climate-change denier? Must they deny that human activities have *any* impact on the climate, or can they merely deny that they have a *significant* impact? What about those who concede that human activities do have a significant impact on the climate but who minimize the seriousness of the effects of climate change? Or those who acknowledge the reality of anthropogenic climate change but claim that we should adapt to it rather than trying to mitigate it? The boundaries of climate-change denial don't appear to be much clearer than those of science denial itself, so even if counter-deniers could offer an exhaustive list of all forms of science denial, it would still be unclear who counts as a science denier, as it would not be clear to whom each entry on the list applies.

---

[41] Specter writes: 'I wish I could say that [the election of President Obama] has helped turn back the greater tide of denialism as well. That would be asking too much. Despite the recession, sales of organic products have continued to grow, propelled by millions who mistakenly think they are doing their part to protect their health and improve the planet. Supplements and vitamins have never been more popular even though a growing stack of evidence suggests that they are almost entirely worthless. Anti-vaccination conspiracy theorists, led by the tireless Jenny McCarthy, continue to flourish' (Specter 2009, 9).

[42] Other counter-deniers who adopt strikingly broad interpretations of the concept (though not as broad as Specter's) include philosopher Adrian Bardon (see, e.g., (Bardon 2019, 49)).

Finally (and, to my mind, most importantly), unlike 'chair', 'science denier' is a term that has a strong evaluative and expressive content and, when such terms do not also have a clear descriptive content, they can be easily abused, and their abuse can itself be epistemically and practically harmful.[43] Let me mention two examples of these harms. First, labels such as 'science denier' can be used to *depoliticize* political debates by dismissing our opponents' opinions and concerns without properly engaging with them. For example, people concerned about the health risks of water fluoridation are often dismissed as science deniers and they are sometimes likened to General Jack Ripper, the paranoid character in Stanley Kubrick's 1964 movie *Dr Strangelove* who believed water fluoridation to be a communist plot. However, a recent systematic review by the US National Toxicology Program suggests that concerns about water fluoridation may not be wholly unfounded.[44] As its critics often point out, liberalism has a tendency to depoliticize political questions by portraying them as technical questions to be answered by experts.[45] However, questions about water fluoridation policies are not merely technical, as they are partly questions about our preferences and values (such as the benefits and the risks of various water fluoridation policies). Moreover, branding certain concerns as science denial might have a chilling effect on scientific research, as researchers might fear that investigating them might negatively affect their professional reputation. The wholesale dismissal of concerns about water fluoridation as science denial, thus, is not only unwarranted given the evidence, but it is also risks stifling scientific inquiry, undermining policy debate, and delaying public policy action.[46]

Second, terms such as 'science denial' can be used to *improperly politicize* scientific questions.[47] For instance, those labelled as science deniers often insinuate that the real science deniers are their opponents. In a 2015 interview, US Senator Ted Cruz, for example, claimed:

> In the debate over global warming, far too often politicians in Washington – and for that matter, a number of scientists receiving large government grants – disregard the science and data and instead push political ideology. (NPR 2015)

Cruz's accusations mirror counter-deniers' own charges and, absent a clear account of science denial, it is unclear how to prevent such abuses of the label.

---

[43] This argument is analogous to one made by Joshua Habgood-Coote (2019) against terms such as 'fake news' and 'post-truth' made.

[44] In particular, it found that exposure to fluoride is inversely associated with IQ scores in children (Taylor et al. 2025).

[45] See, e.g., (Mouffe 2005). [46] See, e.g., (Till and Green 2021).

[47] For two accounts of where to draw the line between proper and improper politicization, see (Aytaç 2021) and (Douglas 2023).

These considerations suggest that, while we might not necessarily need a full-fledged definition of 'science denial' (or an account of science denial in terms of necessary and sufficient conditions), we still must gain a better understanding of the notion of science denial in order to apply it in a consistent and principled manner, to better understand the nature of the phenomenon it is meant to capture, or to assess the validity of the concerns it raises.

In this section, I argue that, once we do so, it becomes apparent that 'science denial' is a highly ambiguous term whose indiscriminate use conflates a heterogeneous set of distinct socio-epistemic phenomena. The more modest goal of this section is to clearly distinguish among some of these specific phenomena and to try to identify among them the one that is most likely to give raise to the counter-deniers' concerns about science denial. Its more ambitious goal is to expose the inadequacy of a number of seemingly plausible 'thin' accounts of science denial – that is, accounts of science denial that describe the phenomenon without appealing to its explanation. In particular, this section argues that no 'thin' account of science denial can differentiate between science denial and several completely distinct phenomena while vindicating both the epistemic and the practical concerns of counter-deniers.

2.2 Science Non-Belief and Science Contestation

One of the most fundamental ambiguities of the term 'science denial' stems from two possible understandings of 'denial' in ordinary English. On the first understanding, 'denial' refers to a type of speech act – the act of denying that something or other is the case. For example, by uttering the words, 'I have never met the victim', the suspect performs the speech act of *denying* having ever met the victim. According to what we might call the *expressive conception* of science denial, science denial consists in the performance of certain speech acts – speech acts that deny the truth of the relevant scientific claims or, more generally, question their scientific standing (e.g., how well-supported they are by the evidence or how widely accepted they are in the relevant scientific community).[48] To avoid confusion, in the rest of this Element, I refer to the performance of these kinds of speech acts in a given context as *science contestation*. So, for example, Donald Trump performed an act of science contestation by declaring on Twitter that 'The concept of global warming was created by and for the Chinese in order to make U.S. manufacturing non-competitive' (Donald J. Trump [@realDonaldTrump] 2012).[49] The expressive

---

[48] See, e.g., (Hoofnagle and Hoofnagle 2007) and (Kahn-Harris 2018).
[49] This is arguably still true even if Trump intended his remarks to be a joke (as he later claimed (see, e.g., Rushing and Martosko 2016))). Presumably, the mere fact that a speaker intends an utterance to be the telling of a joke is not sufficient to turn it into one. Minimally, the speaker

conception of science denial suggests a first possible account of science denial – one according to which a science denier is a *science contester* (i.e., someone who occasionally or systematically performs acts of science contestation in certain contexts).

On the second understanding, 'denial' refers to a psychological state – the state that is colloquially referred to as 'being in denial'.[50] Suppose, for example, that Andy consistently disregards the clear signs of his dependence on alcohol and the problems it causes him and those around him. As a result, Andy believes that he drinks only occasionally and that he could quit drinking anytime if he wanted to. Andy seems to be in a state of denial about his drinking habits. He does not believe that he has a drinking problem despite the fact that the evidence available to him supports the belief that he does.[51] According to what we might call the *epistemic conception* of science denial, science deniers are, like Andy, in denial about inconvenient truths.[52] They do not believe certain scientific claims despite the fact that the evidence available to them supports believing them. The epistemic conception of science denial suggests a second possible account of science denial – namely, one that identifies science denial with *science non-belief*, where, to a first approximation, a *science non-believer* is someone who does not believe certain scientific claims. On this account, someone who does not believe the relevant scientific claims qualifies as a science denier regardless of whether they disbelieve those claims (i.e., believe their negation) or whether they believe any related pseudo-scientific claims (such as the claim that the Earth is flat or that 5G cellphone towers cause COVID-19). However, those who actively disbelieve the relevant scientific claims or who believe pseudo-scientific claims automatically qualify as science non-believers, as those beliefs require not believing the corresponding scientific claims.

These two accounts of science denial disagree not only about the descriptive content of the term 'science denier' but also about its extension. While most science non-believers might also be science contesters, and vice versa, science non-belief and science contestation do not always go hand in hand. On the one hand, someone can be an *insincere (science) contester* – that is, a science

---

must also reasonably expect their audience to be able to recognize their intention (from, e.g., the content of the utterance or the context in which it is made). Arguably, nothing about Trump's tweet indicated that it was meant as a joke.

[50] See, e.g., (Specter 2009, 3) and (Bardon 2019, 1–2).

[51] Throughout this Element, I take someone's evidence to be the total evidence to which they have access (including any evidence to which they access but they ignore or neglect). This Element adopts a broadly evidentialist account of epistemic justification (see, e.g., (Conee and Feldman 2004)), as this seems to be the approach implicitly adopted by most counter-deniers.

[52] It is no coincidence that *An Inconvenient Truth* is the title of the popular documentary about anthropogenic climate change written by former US Vice-President Al Gore.

contester who is not also a science non-believer. For instance, a politician who believes in anthropogenic climate change but claims not to believe in it for political gain would classify as an insincere contester. On the other hand, someone can be a *tacit (science) non-believer* – that is, a science non-believer who is not also a science contester. For example, someone who secretly harbours doubts about the safety of childhood vaccines but would never openly express them for fear of being branded an 'anti-vaxxer' would classify as a tacit non-believer.

Once science non-belief is distinguished from science contestation, it seems clear that, of these two notions, science non-belief is the more likely candidate for a disambiguation of 'science denial' as understood by counter-deniers. Unlike science contestation, science non-belief raises the most serious concern about science denial – namely, the concern that science denial might be directly practically harmful. The point can be seen most clearly by focusing on the 'pure' cases of each phenomenon – that is, tacit non-belief and insincere contestation. The practical concern applies directly to tacit non-belief but only indirectly to insincere contestation. The failure to believe in certain scientific claims might result in tacit science non-believers making decisions that have practically harmful consequences for themselves, for those around them, for their community, or even for the entire world. For example, to the extent that Ben's failure to believe in the reality of anthropogenic climate change contributes to his decision to drive a highly fuel-inefficient vehicle or to oppose policies aimed at mitigating climate change, it seems to have practically harmful consequences for him and everyone else on the planet.[53]

Insincere contestation, on the other hand, does not seem directly practically harmful.[54] Of course, this is not to deny that insincere contestation can be harmful. Science contestation is epistemically harmful to the extent that it pollutes the epistemic environment with misinformation that contributes to disseminating and sustaining science non-belief,[55] and it is indirectly practically harmful to the extent that it contributes to disseminating or sustaining forms of science non-beliefs that are directly practically harmful. The literature on the anti-vaccine movement, for example, suggests that what I call science contestation pollutes the epistemic environment, thereby contributing to disseminating and sustaining

---

[53] I assume that, while Ben's contributions to anthropogenic climate change are likely to be exceedingly small in the grand scheme of things, they are still practically harmful to the extent that they contribute to anthropogenic climate change (see, e.g., (Hiller 2011)).

[54] For the distinction between direct and indirect practical harms, see Section 1.2.

[55] I use 'misinformation' to mean information that is likely to mislead some of its consumers whether deliberately or not. I take misinformation to include disinformation, which is misinformation that is produced to deliberately mislead (some of its) consumers. For a discussion, see (Ridder 2024, 2–4).

unfounded and practically harmful beliefs about vaccines.[56] Similarly, the literature on climate-change denial strongly suggests that what I have called science contestation (and, in particular, insincere contestation) contributes to disseminating and sustaining unfounded beliefs about climate change.[57]

However, the concerns about science contestation ultimately seem to boil down to practical concerns about science non-belief. More specifically, science contestation seems to be concerning only insofar as either it is a symptom of science non-belief (in its sincere form) or it contributes to disseminating and sustaining science non-belief (in both its sincere and insincere forms). Neither of these concerns, however, is about science contestation per se. Both are ultimately concerns about science non-belief. The arguments in Section 1 suggest that, if all forms of science non-belief were completely practically harmless, science contestation would not be an apt object of concern, even if it were a symptom of science non-belief or contributed to it. These considerations suggest that concerns about science denial are primarily concerns about science non-belief and only derivatively concerns about science contestation – science contestation is concerning either as a symptom of science non-belief or as a potential cause of it.

2.3 The Evidence Criterion and the Acceptance Criterion

On my preliminary definition, a science non-believer is someone who fails to believe specific scientific claims. This definition, however, leaves unanswered a crucial question – which scientific claims must one not believe to be classified as a science denier? Unless the class of relevant scientific claims is adequately restricted, the resultant account of science denial risks being overly inclusive, as all of us fail to believe some scientific claims, as we are all partake in what I have called 'ordinary scientific ignorance'. So, which scientific claims must one fail to believe to qualify as a science denier?

Counter-deniers typically appeal to either of two general kinds of criteria to identify the relevant scientific claims. The first kind of criterion focuses on the first-order evidence for the scientific claims. One such criterion is that the relevant scientific claims are *well corroborated* (i.e., strongly supported by first-order empirical evidence).[58] The second kind of criterion focuses on higher-order evidence for those claims. One such criterion is that the relevant scientific claims are *widely accepted* (i.e., accepted by a large proportion of the members of the

---

[56] See, e.g., (Mnookin 2012), (Offit 2015), and (Berman 2020).
[57] See, e.g., (Oreskes and Conway 2010), (Mann and Toles 2016), and (Stott 2021).
[58] See, e.g., (McIntyre 2019, 150).

relevant scientific communities).[59] In what follows, I refer to the first criterion as *the evidence criterion* and to the second one as *the acceptance criterion*.[60]

To my knowledge, counter-deniers never discuss either criterion in detail or explain their reasons for adopting one criterion rather than the other. In fact, some counter-deniers seem to use the two criteria interchangeably.[61] This might be because they assume that the two criteria are always jointly satisfied – that is, that all well-corroborated scientific claims are widely accepted and vice versa. However, this assumption is false. For one thing, in some cases, scientific communities seem to accept claims based on insufficient evidence; and for another, scientific communities sometimes fail to accept a scientific claim despite its being well-corroborated. The case of the aetiology of peptic ulcers can be used to illustrate both points.[62] Until the late 1980s, the biomedical community widely accepted the hypothesis that peptic ulcers were caused by lifestyle factors (such as diet and stress). However, there was little evidence to support that level of scientific consensus.[63] In the 1980s, two Australian physicians, Barry Marshall and Robin Warren, produced increasingly strong evidence that peptic ulcers were in fact caused by a bacterium called *H. pylori*.[64] The biomedical community, however, initially ignored their findings. Increasingly frustrated, Marshall eventually decided to infect himself with *H. pylori*, developed peptic ulcers, and successfully treated them with antibiotics. It was only then that the scientific community started to take the bacterial hypothesis seriously and, eventually, accepted it. In 2005, Marshall and Warren were awarded the Nobel Prize in Medicine for their discovery.

As this case illustrates, the evidence criterion and the acceptance criterion can pull in different directions. For one thing, a scientific community can widely accept a scientific claim even when there is insufficient evidence for it (as the biomedical community did with respect to the lifestyle hypothesis) and, for another, it can fail to widely accept a scientific claim even when there is strong evidence for it (as it did with respect to the bacterial hypothesis). More

---

[59] See, e.g., (Sinatra and Hofer 2021, 9).
[60] For the sake of definiteness, I focus on specific criteria of each kind, as I don't believe that this affects the generality of my arguments.
[61] See, e.g., (Bardon 2019, 14).
[62] See, e.g., (Thagard 1999, chap. 3). While the accuracy of Marshall's version of the story has been questioned (see, e.g., (Levy and Varley forthcoming)), I cannot discuss the issue here. Readers who doubt the accuracy of the peptic ulcer example can replace it with other examples that have the relevant features. For example, the scientific consensus on the origins of COVID-19 was initially that the virus had natural origin. The lab leak hypothesis was widely dismissed as a conspiracy theory. However, that consensus was largely based on two influential letters (one published in *Nature Medicine* and the other in the *Lancet*) whose scientific credentials appear now highly dubious (see, e.g., (Tufekci 2025)).
[63] See, e.g., (Schulz 2010).   [64] See, in particular, (Marshall and Warren 1984).

importantly for us, this case shows that the evidence criterion and the acceptance criterion can disagree on who qualifies as a science denier. Someone who did not believe in the lifestyle hypothesis in the mid-1980s, for example, would count as a science denier according to the acceptance criterion but not according to the evidence criterion. So, which criterion should counter-deniers adopt?

The evidence criterion runs immediately into a number of serious problems. Here, I focus on two of them. The first is that it seems unable to distinguish between science denial and ordinary scientific ignorance. If a science denier is simply someone who does not believe some well-corroborated scientific claims, then we are all science deniers because each of us fails to believe a multitude of well-corroborated scientific claims. For example, presumably, most people do not believe that boron has two stable isotopes or that Arcturus is a red giant even if both claims are well corroborated. Most of us simply do not know that these claims are well corroborated. However, while we might all be scientifically ignorant to some extent, we are not all science deniers.[65] For one thing, if we were all science deniers, then 'science denier' would be a completely useless label, as it would apply to everyone. For another, ordinary scientific ignorance seems epistemically blameless. Given the sheer size of our current scientific knowledge, nobody (not even the most knowledgeable scientist) can be expected to believe every well-corroborated scientific claim in every scientific field.

The second problem is that it seems highly unlikely that most non-experts would be capable of applying the evidence criterion on their own. Only the relevant experts seem capable of gathering and assessing the first-order evidence that bears on each scientific claim. For example, I would never be able to gather and assess first-order evidence that bears on the claim that boron has only two stable isotopes all on my own. For a non-expert like me, the best strategy to determine which scientific claims meet the evidence criterion would be to rely on the opinions of the relevant experts. However, this suggests that the only way for non-experts to comply as closely as possible with the evidence criterion is to employ a criterion such as the acceptance criterion – that is, to use the level of consensus among the experts on a given claim as higher-order evidence of the strength of the first-order evidence for that claim.

The acceptance criterion, however, also faces a number of serious problems. First, it is unclear how widely accepted a claim must be to meet the acceptance criterion. Does it have to be accepted by a simple majority of the community (50 per cent plus one) or by a supermajority (e.g., 95 per cent)?[66] Second, it does

---

[65] Unsurprisingly, counter-deniers agree that science denial should not be confused with ordinary scientific ignorance. See, e.g., (Bardon 2019, 4), (McIntyre 2021, 50–57), and (Sinatra and Hofer 2021, 54–55).

[66] See, e.g., (Vickers 2023).

not seem reasonable to expect non-experts to be able to apply the acceptance criterion either, as, for someone who is not familiar with the relevant scientific community, it is not easy to accurately assess how widely a given scientific claim is accepted in that community (or even identify the relevant scientific community).[67] Third, the acceptance criterion too runs into the problem of distinguishing science denial from ordinary scientific ignorance. If a science denier is simply someone who does not believe a widely accepted scientific claim, then we are all science deniers because each of us fails to believe a multitude of widely accepted scientific claims (such as the claim about the stable isotopes of boron) out of epistemically blameless ignorance.

Fourth, the acceptance criterion would classify all scientific dissenters as science deniers, which seems particularly counterintuitive in the case of scientific mavericks whose unconventional scientific beliefs were eventually vindicated (such as Galileo, Darwin, or the aforementioned Marshall and Warren).[68] In response to this problem, supporters of the acceptance criterion might argue that their account of science denial is only meant to apply to non-experts. This restricted account, however, seems to be inconsistent with the standard use of the label where it applies also to rogue experts such as Andrew Wakefield, the former physician who co-authored the infamous paper suggesting that the MMR vaccine might cause autism and who is now a prominent anti-vaccine activist. Another possible response is to adopt an account of science denial that applies the acceptance criterion to non-experts and the evidence criterion to experts. On this disjunctive account, a science denier is either an expert (in a relevant field) who does not believe some well-corroborated scientific claim or a non-expert who does not believe some widely accepted scientific claim. According to this account, a scientific dissenter qualifies a science denier only if they fail to believe well-corroborated scientific claims (within their area of expertise). This account might include scientific outcasts such as Wakefield

---

[67] This problem is well illustrated by Peter Vickers' efforts to establish whether a 95 per cent scientific consensus has been reached over the cause(s) of the extinction of dinosaurs (Vickers 2023, chap. 7). It seems unreasonable to expect most ordinary people to have the resources (or the motivation) to determine whether a suitable level of consensus has been reached over each and every scientific claim that is relevant to their practical decisions.

[68] It is possible to resist the assumptions that underlie this objection. One might argue that, even if some celebrated scientists were scientific dissenters whose ideas were eventually vindicated, most scientific dissenters are wrong. Additionally, one might argue that the story of the lonesome scientific maverick pushing against the dogmatic scientific consensus of their time is often a myth, as, typically, there was either no scientific consensus at the time or the alleged maverick's views were not resisted and ostracized as stubbornly as the myth suggests (see, e.g., (Vickers 2023)). However, even if both these points are correct, it would still seem inappropriate to brand all scientific dissenters as science deniers, especially in those cases in which the dissenters' beliefs were based on enough evidence, and they eventually become widely accepted.

among the science deniers while clearing scientific mavericks (such as Marshall and Warren) from the charge of science denial. However, it is not entirely clear that the disjunctive account is fully satisfactory. For one thing, it is unclear whether the beliefs of celebrated scientific mavericks are always well corroborated.[69] For another, it is doubtful that experts can base their scientific beliefs exclusively on first-order evidence and not at all on the opinions of their scientific peers.[70] Whatever the right account might be, these brief remarks suggest that the question of what criteria should be used to classify experts as science deniers may require a very different answer from the corresponding question about non-experts. Due to lack of space, I shall set the first question aside in the rest of this Element. In other words, the primary goal going is to identify an account of science denial that applies to non-experts but not necessarily to experts. Such an account might have to be combined with a separate account that applies to experts.

Fifth (and finally), it is dubious that any non-expert who fails to believe a widely accepted scientific claim should be classified as a science denier. Consider for example the case of Carla, a patient with peptic ulcers, who, in the mid-1980s, heard from her doctor about Marshall and Warren's recently published research. Her doctor explained to her that, while the lifestyle hypothesis was widely accepted, two researchers had found credible evidence that a bacterium caused peptic ulcers and that they could be effectively treated with antibiotics. In light of this, Carla agreed to undergo a course of antibiotics, which successfully treated her ulcers. As a result, Carla came to disbelieve the lifestyle hypothesis while it was still widely accepted and believe the bacterial hypothesis instead. The acceptance criterion seems to entail that Carla was a science denier, that she remained one until the scientific community eventually caught up to her, and that, after that happened, she no longer was a science denier. In Carla's case, however, these claims seem counterintuitive, as not only does her case not seem to be a case of epistemically blameworthy scientific ignorance – it does not seem to be a case of ignorance at all.

Hard-line supporters of the acceptance criterion might insist that this is wrong and that the epistemically responsible thing for non-experts is to believe all and only scientific claims that are widely accepted.[71] They might claim that, while Carla's beliefs about peptic ulcers turned out to be true, it was epistemically

---

[69] See, e.g., the classic discussion of Kepler's extra-scientific reason for believing the heliocentric hypothesis in (Kuhn 1957).

[70] According to conciliationism, for example, scientists should also consider higher-order evidence provided by the opinions of their epistemic peers. For the view that even scientists should consider this kind of higher-order evidence, see (Levy and Varley forthcoming); for an overview of the peer disagreement literature, see (Frances and Matheson 2024).

[71] See, e.g., (Anderson 2011).

irresponsible of Carla to disbelieve the lifestyle hypothesis while it was still widely accepted or to believe the bacterial hypothesis before it was widely accepted. The hard-line position, however, seems highly implausible. First, Carla's beliefs did not have to be epistemically irresponsible. Suppose, for example, that Carla's doctor had good reasons to believe that peptic ulcers were caused by *H. pylori* on the basis of Marshall and Warren's research, that she explained those reasons to Carla as clearly and accurately as possible, and that Carla had good reasons to trust her doctor. Under those circumstances, Carla's disbelief in the lifestyle hypothesis and her belief in the bacterial hypothesis would not seem epistemically irresponsible. In fact, it is usually assumed that non-expert should believe the testimony of trustworthy experts, and this is exactly what Carla did.[72]

Second, it seems unreasonable to insist that ordinary people should wait for a consensus to emerge to believe (or act upon) certain scientific claims. If, for example, there is significant evidence that gas stoves cause asthma in children,[73] then this seems to be a sufficient reason for a parent to decide not to buy a gas stove even if there is still no scientific consensus over the claim that gas stoves cause asthma in children.[74] In fact, such a parent can't afford the luxury of waiting for a consensus to emerge before making their decision about whether they should replace their gas stove with an electric one.

Third, the hard-liners' epistemic standards seem too demanding. If anyone who does not believe any widely accepted scientific claim is epistemically irresponsible, then we are all epistemically irresponsible because all of us fail to believe many widely accepted scientific claims out of ordinary scientific ignorance.

Moreover, the hard-line position seems to be even less tenable when applied to other cases. For instance, for much of the nineteenth and twentieth centuries, a broad scientific consensus held that upper-class white men were naturally endowed with greater intelligence than other groups, including women, racial and ethnic minorities, and lower-class individuals.[75] According to hard-liners, those who rejected some of these widely accepted claims – including many first-wave feminists in the late nineteenth and early twentieth centuries – would

---

[72] The interaction between Carla and her doctor, for example, seems to conform to the normative account of scientific testimony defended by Mikkel Gerken (2022). According to Gerken, 'scientific expert testifiers should, whenever feasible, include appropriate aspects of the nature and strength of scientific justification ... for the scientific hypothesis in question' (Gerken 2022, 158), and Carla's doctor did just that in the case discussed. There would be no point in including appropriate aspects of the nature and strength of scientific justification for the bacterial hypothesis if, as a lay person, Carla was completely unable to appreciate the evidence for a scientific hypothesis even when it is properly explained to her.

[73] See, e.g., (Lin, Brunekreef, and Gehring 2013).   [74] See, e.g., (Li et al. 2023).

[75] See, e.g., (Gould 2006).

qualify as science deniers. Yet, if science denial is epistemically blameworthy (or the result of epistemic irresponsibility), this seems to be a misdiagnosis. If anything, it was the scientists of that era, not the first-wave feminists, who were epistemically irresponsible and, arguably, blameworthy. In fact, such cases suggest not only that the failure to believe widely accepted scientific claims can be epistemically blameless (as with ordinary scientific ignorance) but that it may even be epistemically praiseworthy in some cases. Unlike the beliefs of the majority of scientists at the time, the beliefs of first-wave feminists were not only true but, arguably, also epistemically justified.[76] If so, then first-wave feminists were *justified (science) non-believers* – they were epistemically justified in their disbelief of some widely accepted scientific claims.

Similar considerations apply to the case of lay experts (i.e., people who lack formal training or credentials in a given scientific field but who have developed substantial knowledge about it through lived experience, traditional culture, or practical engagement with its subject matter). Hard-liners would either have to deny the existence of lay experts (which is widely accepted among expertise scholars)[77] or maintain that lay experts should always defer to credentialed experts (which is implausible, as it is not unusual for lay experts turn out to be right in cases of disagreement).[78]

The problems with the hard-line position might be avoided by modifying the acceptance criterion so as to exclude cases in which the scientific consensus is based on inadequate or spurious grounds (such as the racist, sexist, and classist prejudices of the relevant scientists). However, it is unclear that any general set of criteria can be used to identify all cases in which the scientific consensus is defective.[79] Moreover, even if such criteria existed, it is dubious that ordinary people would be able to apply them effectively to real-world cases on their own.[80] Finally, any such set of criteria could be misused to strategically disqualify genuine cases of scientific consensus in order to promote certain forms of science non-belief.

---

[76] I take it that first-wave feminists had ample first-hand (first-order) evidence to disbelieve claims to the effect that women were intellectually inferior to men. While this evidence was largely anecdotal, it was arguably stronger than any of the evidence allegedly supporting those claims (see, e.g., (Saini 2017)).

[77] See, e.g., (Collins and Evans 2007) and (Watson 2021).

[78] The classic example of this is discussed in (Wynne 1996).

[79] For an overview of the difficulties arising from trying to distinguish genuine cases of scientific consensus from spurious ones, see (de Melo-Martín and Intemann 2018).

[80] For example, Boaz Miller (2013) proposes three conditions that might distinguish cases of genuine scientific consensus from spurious consensus, which he calls social calibration, apparent consilience of evidence, and social diversity. However, it seems highly unlikely that most ordinary people would be able to apply these criteria to real-world cases of scientific consensus. For a much more exhaustive discussion, see (de Melo-Martín and Intemann 2018).

2.4 Science Disbelief

Counter-deniers might argue that the problem with the account of science denial as science non-belief is that it is too inclusive. Science denial should be identified with a more specific phenomenon – science disbelief. A *science disbeliever* is a non-expert who actively disbelieves some widely accepted scientific claim. Since all science disbelievers must be science non-believers but science non-believers need not be science disbelievers, this account of science denial is less inclusive and, as such, it might avoid some of the problems faced by the more inclusive account in terms of science non-belief. In particular, counter-deniers might hope that adopting this less inclusive account would allow them to distinguish science denial from ordinary scientific ignorance. After all, while most of us might not believe that boron has two stable isotopes, we do not disbelieve that claim either.

However, this account of science denial too faces some problems. Let me mention two. First, the account still fails to distinguish science denial from some of the other phenomena we have discussed. For one thing, many everyday beliefs are at odds with widely accepted scientific claims. Many people believe that cold weather causes colds, and most people believe that the question of whether any two events are simultaneous has an absolute, objective answer. However, while those beliefs are inconsistent with widely accepted scientific claims, they hardly warrant charges of science denial. For another, first-wave feminists disbelieved widely accepted scientific claims about the intellectual inferiority of women but, as argued earlier, should not be labelled science deniers.

Second, if the main concern of counter-deniers is that science denial is practically harmful, it is unclear why they should focus exclusively on science disbelief when science non-belief can be just as practically harmful (if not more). For example, someone who does not believe the claims that vaccines do not cause autism or that HIV causes AIDS may behave exactly as someone who actively disbelieves those claims. Those who produce scientific misinformation seem to know this all too well.[81] For example, the tobacco industry's disinformation campaign concerning the health effects of second-hand smoking did not aim to persuade the public to *disbelieve* the relevant scientific claims (e.g., that second-hand smoking causes lung cancer). It had the more modest aim of trying to persuade them *not to believe* those claims. As a tobacco industry memo infamously put it, 'doubt is our product'.[82] If the primary concern about science denial is its practical harmfulness, it is unclear why counter-deniers

---

[81] This is often true of disinformation campaigns in general. See, e.g., (Pomerantsev 2020).
[82] See, e.g., (Michaels 2008) and (Oreskes and Conway 2010).

would focus exclusively on the narrower phenomenon of science disbelief rather than on the broader phenomenon of science non-belief, given that the latter is no less practically harmful than the former. In fact, some forms of science disbelief (such as flat-Eartherism) are arguably less practically harmful than some forms of science non-belief (such as certain forms of vaccine hesitancy).

2.5 Science Refusal

In response to these problems, counter-deniers may respond that science denial is not merely a *failure* to believe widely accepted scientific claims – it is a *refusal* to believe them. While all of us fail to believe many such claims, most of us do not actively refuse to believe them. On this account, a science denier is a *science refuser* – that is, a non-expert who refuses to believe some widely accepted scientific claim.[83] Yet, it remains unclear what it means to *refuse* to believe a claim (as opposed to merely failing to believe it).

Earlier, I suggested that the reason why ordinary scientific ignorance may be epistemically blameless is that most of us simply do not know that the relevant scientific claims are widely accepted. This inspires the following proposal. A science refuser is a non-expert who does not believe some widely accepted scientific claim despite knowing that it is widely accepted. However, this proposal encounters at least two difficulties. First, someone might know that a scientific claim is widely accepted and yet have good reasons to suspect that the scientific consensus is defective – for example, unreliable research,[84] or community biases,[85] or other systemic failures.[86] Consider again the case of first-wave feminists who rejected the scientific consensus on the intellectual inferiority of women. While they did refuse to believe claims that, at the time, were widely accepted, labelling them as science deniers seems wrong if the label implies that they were epistemically irresponsible or blameworthy. Second, this proposal excludes many of those who counter-deniers would pre-theoretically classify as science deniers. Empirical evidence suggests that many alleged science deniers do not believe that the relevant scientific claims are widely accepted,[87] which, on this

---

[83] See, e.g., (McIntyre 2019, 150) and (Sinatra and Hofer 2021, 9).

[84] See, e.g., (Angell 2005) and (Brown 2004) for worries about the role of conflicts of interest in medical research and (Stegenga 2018) for a more general scepticism about the reliability of claims about effectiveness of medical treatments.

[85] See, e.g., (Gould 2006), (Fine 2011), (Saini 2017), and (Saini 2020).

[86] As in the case of the unsupported belief that peptic ulcers were caused by lifestyle factors or the case of the origins of COVID-19 (see, e.g., (Tufekci 2025)).

[87] See, e.g., (van der Linden et al. 2015), (van der Linden, Clarke, and Maibach 2015), (Kahan 2015), and (Kahan 2017).

proposal, entails that they are not science deniers after all. This proposal is thus both overly inclusive and overly narrow.

Counter-deniers might respond that, even if some science deniers might not know that the relevant scientific claims are widely accepted, they *should* know. However, this proposal is still vulnerable to the first objection: first-wave feminists *knew* that a scientific consensus existed regarding the intellectual inferiority of women, but (correctly) doubted the grounds of that consensus. Moreover, as argued earlier (Section 2.3), it is unrealistic to expect non-experts to be able to know how widely accepted a scientific claim is. Finally, it is unclear whether this proposal can distinguish science denial from ordinary scientific ignorance. After all, why should non-experts know whether there is a scientific consensus on the causes of climate change but not whether there is one on, say, the causes of the extinction of dinosaurs?

Counter-deniers might reply that, in the case of anthropogenic climate change, the duty is not a purely epistemic one but (partly) a moral or political one. People should know that there is a scientific consensus on the causes of climate change because those scientific claims are relevant to their practical decisions as citizens or as consumers while the causes of the extinction of dinosaurs have no practical relevance for them. However, this answer too faces problems. First, even if there was such a duty to know, it would not seem to apply to scientific claims that are not (directly) relevant to people's practical decisions, such as the claim that Earth is a globe or the claim that it is more than 10,000 years old. Second, the first-wave feminist case suggests that there is no moral or political duty to accept the scientific consensus on morally or politically charged issues. Finally, while many people might know that there is a broad scientific consensus about the reality of anthropogenic climate change, very few people seem to know whether or not there is an equally broad scientific consensus over other crucial practically relevant questions about anthropogenic climate change (such as the existence of so-called 'tipping points' or the likely consequences of climate change under various possible scenarios). If the supposed duty to know is triggered by the practical relevance of that knowledge, then wouldn't these be things all of us have a duty to know? And, if we don't, are we all science refusers?

While the suggestion that science deniers *refuse* (rather than merely fail) to believe widely accepted scientific claims might seem promising, without a clear account of science refusal, it merely replaces the obscure notion of science denial with the no-less-obscure notion of science refusal.

2.6 Epistemically Blameworthy Science Non-Belief

Counter-deniers might hope to avoid all of the problems discussed so far simply by identifying science denial with *(epistemically) blameworthy (science) non-belief*. On this account, a science denier is a non-expert whose failure to believe some widely accepted scientific claim is epistemically blameworthy. At first, the blameworthy non-belief account might seem promising. For one thing, it would seem to solve the problem of distinguishing science denial from epistemically blameless forms of science non-belief (such as ordinary scientific ignorance or justified science non-belief). However, on closer scrutiny, its promises appear illusory.

First, unless the account is complemented by an explicit account of epistemic blameworthiness, it is of little use. For one thing, it provides no independent criterion to determine who counts as a science denier and, without an independent criterion, any classification would seem somewhat circular. Would counter-deniers classify Ben as a science denier only after having determined that his beliefs about climate change are epistemically blameworthy or would they simply assume that his beliefs about climate change are epistemically blameworthy because he does not believe in the reality of anthropogenic climate change? For another, each of us would have to rely on their own intuitions to determine whether a given case of science non-belief is epistemically blameworthy, intuitions that may be neither fully dependable nor universally shared.

Second, ideally, an account of science denial should not simply *state that* science denial is epistemically blameworthy – it should also *explain why* it is epistemically blameworthy. However, this account not only fails to do so – it does not even try. The account seems to suggest that the alleged epistemic blameworthiness of science denial is a mystery that cannot be further explained or investigated.

Finally, some cases of epistemically blameworthy non-belief do not seem to qualify as science denial. Consider, for example, what we might call *indifferent (science) non-belief*. Had I studied chemistry more diligently in high school (or paid better attention in class), I would now believe many widely accepted scientific claims that I do not currently believe. Arguably, my ignorance of these claims is epistemically blameworthy. I would seem worthy of epistemic blame for failing to believe widely accepted scientific claims that I could have easily believed had I taken my epistemic responsibilities more seriously. Yet, while indifferent non-belief seems to be epistemically blameworthy, it clearly does not deserve to be classified as science denial.

2.7 Science Non-Compliance

In light of the problems faced by epistemic accounts of science denial, counter-deniers might consider adopting an account that focuses on the practical behaviour of science deniers rather than on their epistemic behaviour. In particular, counter-deniers might adopt an account according to which a science denier is a *science non-complier* – that is, a non-expert who does not not to follow recommendations or regulations based on widely accepted scientific claims. This approach appears to avoid some of the problems plaguing epistemic accounts. Most importantly, it seems to distinguish science denial from ordinary scientific ignorance. While all of us fail to believe many widely accepted scientific claims, most of us are still willing to follow recommendations or regulations based on those claims, which means that we would not qualify as science deniers under this account.

However, the account is obviously overly inclusive. Science non-compliance can be the result of a multiplicity of factors that seem to have little or nothing to do with science denial. For example, in the United States, uninsured patients have significantly lower vaccination rates than insured ones.[88] This suggests that limited access to healthcare and other socio-economic factors play a significant role in many cases of science non-compliance. In other cases, science non-compliance might be due to prudential considerations. Consider, for example, the case of Daphne, who disbelieves allegations that the MMR vaccine causes autism and who is aware that public health guidelines recommend that children her sons' age should receive their first dose of the MMR vaccine. However, her belief that the MMR vaccine does not cause autism is not strong enough to overcome her fear that her son might develop autism.

For the sake of clarity, it is convenient to frame this phenomenon in decision-theoretic terms. Suppose that Daphne attaches a much greater disutility to the very unlikely (even by her own lights) possibility that her son might develop autism due to the MMR vaccine than to the much more likely possibility that he might contract measles, mumps, or rubella as a result of not being vaccinated against those diseases. If the discrepancy between her utilities is large enough, then the expected utility of not vaccinating her child will exceed the expected utility of vaccinating him even if her credence in in the claim that the MMR vaccine causes autism is much lower than her credence in its negation. In fact, according to expected utility theory, given Daphne's beliefs, and her preferences, it would be rational for her to decide not to vaccinate her child.[89]

---

[88] See, e.g., (Blewett et al. 2008), (Gaffney, Woolhandler, and Himmelstein 2022), and (Gaffney et al. 2023).

[89] It is worth noting that equivalent results could be obtained if Daphne adopts a different decision-making procedure. For example, if, instead of aiming to maximize her expected utility, Daphne aims to maximize the worst outcome, then she will still choose not to vaccinate her child as long as her utilities stay the same.

The key point is that, while Daphne is a science non-complier, she is not a science non-believer, as she does believe that the MMR vaccine does not cause autism. Rather, Daphne is what we might call a *prudential (science) non-complier*. In her case, prudential considerations trump epistemic ones. It should be also noted that, while, in some cases, the preferences of prudential non-compliers might be the result of prejudices (such as ableist prejudices about autism) and other inaccurate factual beliefs (such as beliefs about the seriousness of diseases such as measles, mumps, or rubella), this need not always be the case. In some cases, it might simply be based on a simple discrepancy between the values of the prudential non-compliers and those on which the relevant recommendations or regulations are based.

Crucially, while attempts to correct the beliefs that underpin Daphne's preferences might be effective in some cases of prudential non-compliance, attempts to increase her credence in the claim that the MMR vaccine causes autism might have no effect on her behaviour as long as prudential considerations trump epistemic ones. In the most extreme cases, prudential non-compliance becomes what we might call *prudential swamping* – the disparity in the utilities of the outcomes is so large that the expected utility of non-compliance exceeds that of compliance no matter how high the subject's credence in the relevant scientific claim is (as long as is not one). Again, the only effective approach in cases of prudential swamping might be to address any mistaken factual beliefs underlying the prudential swamper's preferences.

It is worth noting that prudential non-compliance is not a purely theoretical possibility. Some widely discussed empirical studies seem to show that convincing vaccine hesitators of the safety and effectiveness of vaccines does not necessarily increase their intent to be vaccinated (or have their children vaccinated).[90] This sort of result is often interpreted as evidence of the stubbornness of science deniers.[91] However, the notion of prudential non-compliance offers an alternative explanation that does not postulate an epistemic failing on the part of these vaccine refusers – namely, the decisive factor in their decision to refuse vaccines is not their science non-belief, but their prudential-science non-compliance.

Counter-deniers might argue that these problems can be avoided by restricting the account so as to include only non-experts who, unlike Daphne, are both science non-believers and science non-compliers. However, this account still faces a number of problems. First, according to it, prototypical examples of science denial such as flat-Eartherism or creationism do not qualify as cases of science

---

[90] See (Nyhan et al. 2014) and (Nyhan and Reifler 2015).
[91] See, e.g., (Mooney 2014) and (Alter 2014) (see (Goldenberg 2021, chap. 2) for additional references and for an excellent discussion).

denial, as neither system of beliefs requires those who adopt them not to follow any recommendations or regulations based on widely accepted scientific claims.

Second, it is not easy for a non-expert to determine whether a piece of scientific advice is based on scientific claims that are widely accepted. For example, early in the COVID pandemic, the US Centers for Disease Control advised people not to wear face masks unless they were ill. However, for most non-experts at the time, it would have been unclear whether that piece of advice was based on widely accepted scientific claims about the effectiveness of face masks against COVID-19 (rather than on concerns about equipment shortages for healthcare workers). If, as I argued earlier, it is unreasonable to expect non-experts to be able to figure out whether a given scientific claim is widely accepted, it seems even more unreasonable to expect them to be able to figure out which set of scientific claims each piece of scientific advice they receive is based on and whether each of those claims is widely accepted.

Third, there are many cases in which recommendations are based on scientific claims that are widely accepted but false. For example, in the 1980s, the recommendation for Carla would have been not to use antibiotics to treat her peptic ulcers. However, as I argued earlier, it would seem wrong to classify Carla as a science denier even if she did not follow recommendations that were based on scientific claims that were widely accepted at the time.

Finally, no recommendation or regulation is ever based exclusively on scientific claims. Recommendations or regulations always also presuppose value judgments. So, it is possible that a science non-believer who does not follow a piece of scientific advice would still not follow it even if she were to come to believe the widely accepted scientific claims on which it is based. This seems to suggest that, even in those cases in which science non-compliance is partly the result of science non-belief, there is no guarantee that addressing science non-belief would lead to the former science non-believers making less practically harmful decisions, as they might also be prudential non-compliers. As I argue in more detail in the next section, part of the challenge in assessing the validity of the counter-deniers' practical concerns is that it is often extremely difficult to determine whether or not science non-belief plays any role in someone's making practically harmful decisions and, even more so, whether or not it plays a *decisive* role – that is, whether the science non-believers would make different practical decisions if it were not for their science non-belief.

2.8 Conclusion

This section explored several 'thin' accounts of science denial and found them all inadequate. While, admittedly, these arguments fall short of demonstrating that no 'thin' account of science denial is tenable, they do seem to show that the task of formulating a 'thin' account of science denial is much harder than counter-deniers seem to think. A much more promising option is to turn instead to 'thick' accounts of science denial – that is, accounts of science denial that describe the phenomenon by explaining its causes. The next section turns to the most promising and prominent account of this kind, which explains science denial in terms of motivated reasoning.

3 What Explains Science Denial?

3.1 Motivated Science Non-Belief

The previous section argued that 'thin' accounts of science denial (i.e., accounts of science denial that attempt to describe the phenomenon without explaining it) face very serious problems. This section turns to what is arguably the most promising and prominent 'thick' account of science denial, which claims that science denial results from motivated reasoning.[92]

Motivated reasoning is 'the tendency of individuals to unconsciously conform assessment of factual information to some goal collateral to assessing its truth' (Kahan 2016, 2). More precisely, someone engages in *motivated reasoning* about a given issue whenever the aim of their inquiry into that issue is to achieve some non-epistemic goals of theirs at the expense of epistemic goals, such as the goals of acquiring true beliefs or avoiding false ones. For the sake of clarity, it is helpful to distinguish *distal goals* from *proximate goals*, where a (more) proximate goal is an intermediate goal that is instrumental to the achievement (more) distal goals and one's ultimate goals are one's most distal goals. When I double-check that my appointment with the dentist is at 3 p.m., my distal goals might be exclusively non-epistemic (e.g., the goal of not being late for my appointment). However, if my inquiry into the claim that my appointment is at 3 p.m. is to contribute to achieving my distal non-epistemic goals, it must also aim to achieve the proximate epistemic goal of having a true belief about the time of my appointment. So, even if all of my distal goals are non-epistemic, I am not engaging in motivated reasoning, as my inquiry does not aim to achieve my distal non-epistemic goals *at the expense of* the proximate epistemic goal of having a true belief about the time of my appointment. In fact, the opposite is true, as I realize that it is only by achieving the proximate epistemic goal of having a true belief about the time of my appointment that I can achieve my distal non-epistemic goal of not missing it. If, on the other hand, my only goal in inquiring about the time of my appointment is the non-epistemic goal of placating my anxiety about being late to my

---

[92] See, e.g., (Specter 2009, 2–3), (Bardon 2019, 4), (McIntyre 2019, 150), and (Sinatra and Hofer 2021, chap. 6).

appointment (regardless of whether my appointment is, in fact, at 3 p.m.), then I would be engaging in motivated reasoning, as my inquiry about that claim would aim to achieve a non-epistemic goal *at the expense of* the epistemic goal of having a true belief about the time of my appointment. Similarly, Andy engages in motivated reasoning about his drinking habits if his inquiry into the topic aims to achieve non-epistemic goals (e.g., the goal of feeling better about himself) at the expense of epistemic goals, such as that of avoiding false beliefs about his drinking habits.

The *motivational account* of science denial maintains that a science denier is a *motivated (science) non-believer* – that is, someone who fails to believe some widely accepted scientific claim as a result of motivated reasoning.[93] On this account, people who do not believe in the reality of anthropogenic climate change or the safety of childhood vaccines fail to do so because their reasoning about those issue aims to achieve some non-epistemic goals of theirs (e.g., the goals of protecting their political ideology or that of affirming their social identity) at the expense of any epistemic goals (e.g., the goal of having true beliefs about those issues). As philosopher Lee McIntyre puts it:

> For some people, their primary allegiance is to some sort of ideology. Their beliefs on empirical subjects seem based on fit not with the evidence but rather with their political, religious, or other ideological convictions. When these conflict – and the conclusions of science tread on some sacred topic on

---

[93] According to many counter-deniers, cognitive biases, in general, and confirmation bias, in particular, might also be implicated in science denial (see, e.g., (Bardon 2019, 31) and (Sinatra and Hofer 2021, 14)). The main difference between cognitive biases and motivated reasoning is that, while the former are topic-neutral, the latter is not. So, for example, confirmation bias is (roughly) the tendency to overestimate evidence that confirms one's beliefs and disregard or overestimate evidence that is disconfirms them. However, confirmation bias can affect any of our beliefs regardless of their topic, while motivated reasoning is only supposed to affect beliefs that are more or less directly linked to our social, political, or religious identities (for a discussion of the relevance of this distinction to the broader debate about motivated reasoning, see (Williams 2023a)). The reason why counter-deniers typically focus on motivated reasoning is, presumably, that cognitive biases alone cannot explain some prototypical cases of science denial. For example, while confirmation bias may explain why someone who already believes the Earth to be flat would be unwilling to change their mind despite contrary evidence, it cannot explain how someone who was taught that the Earth is a globe from a very young age would come to reject that belief later in life. In fact, an account relying solely on confirmation bias would predict that those people would be likely to stick to their long-held beliefs about the shape of the Earth and underestimate any evidence that does not support them. Motivated reasoning, thus, seems to offer a much more plausible explanation of the aetiology of flat-Eartherism, as people often seem to embrace flat-Eartherism in the pursuit of non-epistemic goals (such as the goal of feeling special or different). This kind of explanation fits well with first-hand anecdotal evidence from former flat-Earthers (see, e.g., (West 2018, chap. 14)) as well as with more systematic evidence about the adoption of conspiracy theories (see, e.g., (Imhoff and Lamberty 2017)).

which people think that they already know the answer (e.g., whether prayer speeds up healing, whether ESP is possible), this can result in rejection of the scientific attitude. (McIntyre 2019, 151)[94]

The motivational account has several features that counter-deniers are likely to find appealing. Let me mention six here. First, if motivated reasoning is inherently epistemically irresponsible (or blameworthy),[95] then the motivational account can explain why science denial is also epistemically irresponsible (or blameworthy), thus validating the counter-deniers' deontic epistemic concerns.[96] Second, the motivational account can easily distinguish science denial from other forms of science non-belief, such as ordinary scientific ignorance, justified non-belief, and indifferent non-belief. Unlike those other forms of science non-belief, science denial is the result of motivated reasoning. Third, the motivated account seems to apply to experts as much as non-expert. A scientific expert who refuses to believe widely accepted scientific claims due to motivated reasoning is as much a science denier as a non-expert that does the same. Fourth, if, as the Andy example shows, the state of being in denial can be construed as a product of motivated reasoning, the motivational account can also explain in what sense science deniers are in denial. Fifth, the motivational account can explain in what sense science deniers can be regarded as being wilfully ignorant about the relevant scientific claims or as refusing to believe those claims (as opposed to merely failing to believe them). It seems natural to describe someone who does not believe a claim as a result of motivated reasoning in those ways. For example, if Andy does not believe that he has a drinking problem because his reasoning about the matter aims to achieve the non-epistemic goal of not feeling bad about himself at the expense of the epistemic goal of having true beliefs on the matter, then it would be natural to describe him as refusing to believe that he has a drinking problem or as being wilfully ignorant of his drinking problem. Finally, the motivational account supports the post-truth diagnosis of the collective epistemic malaise afflicting liberal democracies. As mentioned earlier (Section 1.3), the post-truth diagnosis attributes that malaise to a lack of interest in (or concern for) the truth (and other

---

[94] See, also, (Bardon 2019, 4), (McIntyre 2019, 150), and (Sinatra and Hofer 2021, chap. 6).

[95] The question of whether motivated reasoning is inherently epistemically blameworthy is a complex one that I will not address here. For an overview of the literature on motivated reasoning and normative epistemology, see (Ellis 2022). Personally, I believe that proponents of inductive risk in philosophy of science (e.g.: (Douglas 2009)) and of pragmatic encroachment in epistemology (e.g.: (Fantl and McGrath 2009)) have convincingly argued that our non-epistemic goals can legitimately affect our epistemic standards in ways that might be indiscernible from motivated reasoning, which means that much of what passes for motivated reasoning is epistemically blameless.

[96] Counter-deniers could, for example, rely on the standard assumption that there is a strong connection between epistemic responsibility and epistemic goals (see, e.g., (BonJour 1978, 5)).

epistemic goals) on the part of a growing number of our fellow citizens. If one of the main symptoms of that malaise (i.e., science denial) turns out to be the result of reasoning that aims to achieve non-epistemic goals at the expense of epistemic ones, then this would provide strong support for the post-truth diagnosis. The motivational account of science denial can thus be seen as a crucial test case for the post-truth diagnosis.

It is important to note, however, that, in order to have all of the aforementioned features, the motivational account must be combined with an account of motivated reasoning that differs significantly from the one typically adopted by cognitive scientists.[97] Cognitive scientists who work on motivated reasoning are primarily interested in studying how non-epistemic factors such as values, social identities, and practical goals affect human cognition. What I shall call the *universalist account* of motivated reasoning views motivated reasoning as a universal, involuntary, and unconscious feature of human cognition. On the universalist account, the beliefs of those who believe widely accepted scientific claims about the reality of anthropogenic climate change or the safety of childhood vaccines are just as likely to be unconsciously influenced by their values, their social identities, or their practical goals as the beliefs of those who do not believe in those scientific claims. The universalist account is both normatively symmetric and normatively neutral. It is *normatively symmetric* because those who believe in the reality of anthropogenic climate change or the safety of childhood vaccines rely on motivated reasoning just as much as those who do not, so the ones are no more epistemically irresponsible or blameworthy than the others, and it is *normatively neutral* because, since motivated reasoning is both unconscious and involuntary feature of human cognition, it would be inappropriate to hold either side responsible for engaging in it or blame them for it.

In order to have the aforementioned features, the motivational account must thus be combined with a *particularist account* of motivated reasoning instead of a universalist one. On the particularist account, motivated reasoning is neither universal nor entirely involuntary or unconscious. It is a kind of reasoning people employ in a not-entirely-involuntary and not-entirely-unconscious endeavour to preserve their own ignorance on certain topics. The particularist

---

[97] For reasons of space, I cannot discuss the empirical literature on motivated reasoning about scientific matters here. However, I should mention that, while some studies support the view that motivated reasoning plays a significant role in our reasoning about scientific matters (see, e.g., (Kahan et al. 2012)), their conclusions are not entirely uncontroversial either (see, e.g., (Druckman and McGrath 2019), (Tappin, Pennycook, and Rand 2020a) and (Tappin, Pennycook, and Rand 2020b)). For a meta-analysis of studies on motivated reasoning, see (Ditto et al. 2019). For an excellent review of the debate, see (Williams 2023a). Williams also suggests a distinction between motivated reasoning and motivated cognition that is somewhat like the one drawn here between the particularist conception and the universalist one.

conception of motivated reasoning is thus much closer to the notion of wilful ignorance employed by philosophers than to the universalist notion of motivated reasoning employed by cognitive scientists. The particularist account holds that people engage in motivated reasoning in an attempt to remain ignorant of truths that are inconvenient to them, which is why their ignorance is wilful. It is the fact that their use of motivated reasoning is neither entirely involuntary nor entirely unconscious that makes the resulting ignorance both wilful and epistemically culpable. On this account, motivated science non-believers employ motivated reasoning to wilfully ignore inconvenient truths such as the reality of climate change, which makes their epistemic conduct epistemically irresponsible (and blameworthy). As these remarks suggest, the motivational account has all the features mentioned earlier only when it is combined with a particularist account of motivated reasoning. It is for this reason that, unless otherwise stated, in what follows, I use 'motivated reasoning' in its particularist sense, not the universalist one. While many of the arguments in this section apply also to those who adopt a universalist account, the main target are counter-deniers, who adopt a particularist one.

This section argues that, although the account of science denial as motivated non-belief fares much better than 'thin' accounts of science denial discussed in the previous section, it does not vindicate all of the concerns of counter-deniers. Once we carefully distinguish motivated non-belief from other superficially similar phenomena, it seems dubious that any of them warrants all the concerns that animate counter-deniers. More specifically, I argue that counter-deniers face a dilemma. On the one hand, even if motivated non-belief might be epistemically irresponsible (or blameworthy), it is doubtful that it is ever practically harmful. On the other hand, even if other forms of science non-belief may be sometimes practically harmful, it is doubtful that they are always (or ever) epistemically irresponsible (or blameworthy).

3.2 Misinformed Science Non-Belief

This subsection argues that there is a significant tension between motivated reasoning and another factor often implicated in the aetiology of science denial – exposure to scientific misinformation. Even counter-deniers typically concede that exposure to scientific misinformation can contribute to the science deniers' failure to believe certain scientific claims.[98] In fact, as we saw in the previous section, one of the main concerns about science contestation is that it can have

---

[98] See, e.g., (McIntyre 2018, chap. 2) and (Bardon 2019, chap. 2). While it might be clear by now, I should mention that I do not include among counter-deniers those who think of science non-belief primarily in terms of exposure to misinformation (see, e.g., (Lewandowsky et al. 2020)).

epistemically harmful consequences for those who are exposed to it – it can turn them into science non-believers.

For our purposes, it is useful to distinguish motivated non-believers (who do not believe widely accepted scientific claims due to motivated reasoning) from *misinformed (science) non-believers* (who do not believe widely accepted scientific claims *exclusively* due to exposure to misinformation). A few remarks are in order here.

First, this distinction does not deny that misinformation plays a role in motivated non-belief as well. However, misinformation plays distinct roles in motivated non-belief and misinformed non-belief. In particular, exposure to misinformation (as understood here) is a passive phenomenon, which must be distinguished from the active pursuit of misinformation often associated with motivated reasoning. If Evan engages in motivated reasoning about climate change to protect his political beliefs, he might actively seek out (whether consciously or not) misinformation that supports his beliefs about climate change. In my terminology, Evan is not exposed to misinformation, as his willingness to trust the sources of that misinformation partly depends on the content of the misinformation they provide to him – he is willing to trust a given source in part *because* it provides him with misinformation that supports his beliefs about climate change. When Evan seeks out misinformation, his encounters with misinformation are a result of motivated reasoning, not accidental exposure. Frank, on the other hand, has been exposed to misinformation about climate change from a variety of sources that he independently trusts, such as his favourite news channel, his favourite radio talk-shows, his favourite politicians, his pastor, his friends, and his family. Crucially, his willingness to trust those sources, unlike Evan's, is (largely) independent of the content of the misinformation they spread. He trusts those sources on independent grounds, not because they spread the kind of misinformation he was looking for as in the case of Evan. For example, Frank might trust his friends because he has independent first-hand evidence of their trustworthiness in other domains, and he trusts his favourite news channel because it enjoys a good reputation among those he trusts. In my terminology, while Frank is exposed to misinformation, Evan is not – he (actively) *seeks out* misinformation rather than being (passively) *exposed* to it.

Second, the fact that Evan is a motivated non-believer (about climate change) does not mean that he cannot be *also* a misinformed non-believer (on the same topic). In fact, given his political beliefs and the social network in which he is embedded, it is likely that Evan already trusts sources that spread misinformation about climate change independently of the fact that they spread misinformation about it. If this is so, then Evan is an 'impure' motivated non-believer – his science non-belief is due to a combination of motivated reasoning and exposure to misinformation. While it is likely that most actual motivated non-believers are 'impure' motivated non-believers,[99] for the sake of clarity, I focus primarily on 'pure' cases here – cases in which motivated non-believers are not at all exposed to misinformation (even if they might actively seek it out and find it).

Third, even if there are strong reasons to suspect that most motivated non-believers in the real world are 'impure' motivated non-believers, there seems to be no reason to think that there are no misinformed non-believers out there.[100] In fact, it seems highly plausible to assume that some people fail to believe widely accepted scientific claims *exclusively* as a result of their exposure to misinformation without motivated reasoning playing any role in their science non-belief. Consider, for example, a trivial case of misinformed non-belief – that of Geeta, a high school student who, due a typo in her chemistry textbook, comes to mistakenly believe that uranium-236 (not uranium-238) is the most abundant naturally occurring isotope of uranium. Clearly, Geeta is a misinformed non-believer, as her false belief about the isotopes of uranium does not serve any non-epistemic goals of hers.[101] However, if Geeta is a misinformed non-believer, then it is not clear why Frank would not also be one. After all, his case differs from hers only in degree, not in kind. While Frank has been exposed to much more misinformation from a much wider variety of trusted sources than Geeta, both are misinformed non-believers.

Once misinformed non-belief is carefully distinguished from motivated non-belief, it should be obvious that misinformed non-believers should not be classified as science deniers. People who do not believe a given claim as a result of their exposure to misinformation are typically not in denial about that claim. For example, if you mistakenly believe that your flight is boarding because the airport monitors erroneously said so, you are not in denial about the fact that your flight is not boarding yet – you have been simply misinformed about the status of your flight. Moreover, many forms of misinformed non-belief seem completely epistemically blameless. Geeta, for example, is a misinformed non-believer, but her failure to believe that uranium-238 (not uranium-236) is the most abundant

---

[99] It is difficult to see how someone like Evan could be a motivated non-believer without also being exposed to misinformation. Presumably, Evan is surrounded by motivated non-believers and, if most motivated non-believers are also science contesters, then this is enough for him to be exposed to misinformation. Perhaps the only exception to this would be people who actively surround themselves with other motivated non-believers as part of their effort to support their beliefs (as opposed to being passively surrounded by them).

[100] For the sake of simplicity, I am relying on a definition of 'misinformed non-belief' that does not admit any 'impure' cases. However, a better definition would be one that classifies someone as a misinformed non-believer if exposure to misinformation is the predominant factor in their science non-belief (as opposed to the only factor). On this classification, it is likely that many 'impure' motivated non-believers would be classified as 'impure' misinformed non-believers.

[101] The case of Geeta shows that misinformed non-belief would be a phenomenon even if the universalist account of motivated reasoning is true.

naturally occurring isotope of uranium does not make her a science denier, as she has no motive to refuse to believe that particular claim or be in denial about it.

As this last example shows, questions about epistemic responsibility bring into sharper focus the tension between these two diagnoses. While motivated non-belief might be always epistemically irresponsible (or blameworthy), it is doubtful that misinformed non-belief is typically epistemically irresponsible (or blameworthy). To see why, consider the analogous case of someone who falls for a scam. While the moral responsibility for scams might always fall squarely on the scammer, apportioning the epistemic blame is a bit more complicated. In the case of someone who falls for a blatant scam (such as a 'Nigerian prince' scam), it might be tempting to attribute some of the epistemic responsibility to the victim, as they could be blamed for being gullible and gullibility is a form of epistemic irresponsibility. However, in the case of a sophisticated scam (such as a scam pulled by a group of con artists who have carefully cultivated their victim's trust over a long period of time), the epistemic responsibility seems to fall exclusively on the scammers, as we recognize that, in such a case, it would be unreasonable to expect the victim to be able to realize that they are falling victim to a scam. From an epistemological point of view, many cases of misinformed non-belief seem much closer to the sophisticated scam than to the blatant one. Frank's case, for example, seems to be much closer to the case of the victim of a sophisticated scam by a group of con artists who have cultivated his trust over an extended period of time than to that of the victim of a blatant scam. If so, then Frank is neither epistemically irresponsible nor worthy of epistemic blame.

Moreover, if Frank is epistemically irresponsible (or blameworthy), it is because he placed his trust in sources that are not trustworthy (at least with regard to climate change). However, while, sometimes, it might be epistemically irresponsible or blameworthy to place one's trust in untrustworthy sources (e.g., when one has reason to doubt their trustworthiness), this is not always the case. Our epistemic environment is increasingly flooded with both information and misinformation and it is practically impossible for any of us to discern one from the other all on our own. When it comes to information such as scientific information that we have no way to independently verify on our own, we seem to have no other realistic option but to trust the sources that have a good reputation within our own community and mistrust those that do not (just like Frank does).[102] This makes us epistemically dependent on our own communities. If we are fortunate enough to inhabit an epistemically discerning community (i.e., one in which the reputation of various sources of information reliably

---

[102] I sketched this kind of communitarian view in (Contessa 2023), and I develop it further in (Contessa MS). It is worth noting that this kind of view can explain why beliefs on scientific matters are often polarized along ideological lines without appealing to motivated reasoning.

tracks their actual trustworthiness on each of the topics they cover), then we tend to trust trustworthy sources and mistrust untrustworthy ones.[103] If, on the other hand, we have the misfortune to inhabit a community that is not epistemically discerning, then we might trust untrustworthy sources and mistrust trustworthy ones. The crucial difference between the good case and the bad one seems to be a difference in epistemic luck, not epistemic responsibility. It would seem unreasonable to expect Frank to mistrust sources that are well reputed in his community and trust ones that are ill-reputed unless he has independent reason to doubt that his community is epistemically discerning, and it is at best unclear that Frank could ever have any such reason.[104]

Finally, even if, contrary to the previous argument, it was epistemically irresponsible for Frank to place his trust in those sources, the epistemic irresponsibility involved would be different in both degree and kind from the epistemic irresponsibility that is allegedly involved in motivated non-belief. At most, a misinformed non-believer like Frank has chosen the wrong means to achieve his epistemic goals; a motivated non-believer like Evan, on the other hand, is not trying to achieve any epistemic goals in the first place.[105]

These arguments show that, while, in the real world, it might not always be easy to discern misinformed non-belief from motivated non-belief, we must carefully distinguish the one from the other if we want to better understand (or address) either phenomenon, as there are crucial differences between them. On closer scrutiny, cases of misinformed non-belief seem much closer to cases of ordinary scientific ignorance than to cases of science denial (whether understood as motivated non-belief or in the pre-theoretical sense).

---

[103] Personally, I am sympathetic to an internalist communitarian account according to which an epistemically discerning community is one that, collectively, has good reasons to believe that some sources are trustworthy while others are not. However, this idea can be taken in an externalist direction analogous to the one developed by Sandy Goldberg (see, e.g., (Goldberg 2010) and (Goldberg 2018)).

[104] Frank could only have independent reasons to doubt the epistemic discernment of his own community if he could assess the trustworthiness of a variety of sources of information all on his own. However, it is unclear whether Frank would be able to do so. For a discussion of this issue (see, e.g., (Nguyen 2020)).

[105] My diagnosis in cases of misinformed non-belief is somewhat similar to Neil Levy's diagnosis of what he calls denialism, which is that deniers and non-deniers adopt different patterns of epistemic deference (see, in particular, (Levy 2017)). There are, however, a number of significant differences between our views. Let me focus on the three crucial differences. First, insofar as Levy and I reach a similar conclusion, we reach it from different sets of premises. Our arguments are, thus, largely complementary. Second, unlike me, Levy seems to agree with counter-deniers in adopting a monistic account of science denial. I, on the other hand, doubt that all alleged cases of science denial fit the same diagnosis. Third, Levy and I seem to disagree even on which cases are most amenable to a diagnosis of misinformed non-belief. Unfortunately, for reasons of space, I cannot discuss Levy's views more extensively here.

3.3 Misinformed Non-Belief in High-Stakes Cases

Once misinformed non-belief is carefully distinguished from motivated non-belief, it is easier to see why the counter-deniers' practical concerns may be largely misplaced. It is because misinformed non-belief is a much more plausible diagnosis than motivated non-belief in high-stakes cases of alleged science denial – that is, cases in which science non-belief contributes to decisions that have serious practically harmful consequences for the science non-believer or their loved ones.

Consider, for example, the case of Helen, an apprehensive first-time parent who has concerns about vaccinating her daughter. After her concerns are hastily dismissed by her overworked family doctor, Helen researches the topic online and stumbles upon a support group for parents who believe their children have developed autism as a result of vaccines. The parents in the group, who, unlike her doctor, seem to genuinely care about the well-being of Helen's daughter, tell Helen about how their children were diagnosed with autism soon after receiving their vaccinations and point her to a number of online resources that claim that childhood vaccines can cause autism and several other serious conditions. As a result, Helen comes to believe that childhood vaccines cause autism and decides not to have her daughter vaccinated.[106]

It is highly implausible to suggest that Helen is a motivated non-believer. In fact, we can plausibly assume that Helen's ultimate goal in deciding whether or not to believe that vaccines cause autism is to keep her daughter safe and healthy; and, while, admittedly, that's not an epistemic goal, we can also plausibly assume that Helen realizes that, in order to pursue it, she must pursue proximate epistemic goals such as having true beliefs about the safety of vaccines that are instrumental to its achievement. This seems to suggest that, if Helen's conduct is epistemically irresponsible, it is not because her reasoning about the safety of vaccines aims to achieve her non-epistemic goals *at the expense of* epistemic ones. Not only her distal non-epistemic goals are compatible with her epistemic goals, but Helen realizes that her epistemic goals are instrumental to her non-epistemic goals. Nor is it because Helen actively sought out misinformation about the topic, as Helen stumbled into the online support group accidentally and trusted them more than her doctor because, unlike him, they genuinely seemed to care about the well-being of her child. If Helen's conduct is epistemically irresponsible, it is because she has employed epistemic means that are inadequate to the achievement of her epistemic goals. She tried to

---

[106] It is worth noting that exposure to misinformation does not need to turn Helen into a science non-believer to be practically harmful. All it takes is that Helen is exposed to enough misinformation for her to become a prudential non-complier.

assess the truth of the claim that vaccines cause autism by relying on sources that are not trustworthy on that topic.[107] But this means that Helen is a *misinformed* non-believer, not a motivated non-believer.

Moreover, it is not even clear what non-epistemic goals Helen would achieve by believing that vaccines cause autism. If motivated reasoning is primarily used to protect one's wilful ignorance of inconvenient truths (as in Andy's case), then it is implausible to suggest that Helen is engaging in motivated reasoning about the safety of vaccines or that she is in denial about it. The suggestion that Helen is in denial would make much more sense if there was no safe and effective way to protect her daughter from a serious illness but, nevertheless, Helen believed that she could do so by performing a religious ritual or giving her homeopathic remedies. However, it is unclear why Helen would be in denial about the existence of a safe and effective way to protect her daughter from several serious illnesses, as, if anything, that is the opposite of an inconvenient truth.

While not all vaccine-hesitant parents might have the best interest of their children at heart, presumably, most of them do. If so, then it is plausible to assume that most parents who do not believe that childhood vaccines are safe are misinformed non-believers like Helen rather than motivated non-believers. If the ultimate goal of most vaccine-hesitant parents is keeping their children safe and healthy, they presumably realize that, in order to achieve that goal, they must pursue epistemic goals that are instrumental to it, such as the goals of acquiring true beliefs about the safety of vaccines and avoiding false ones. In turn, this means that it is highly implausible to assume that they engage in motivated reasoning about the safety of vaccines. It is much more plausible to assume that, like Helen, those parents are misinformed non-believers – their mistaken beliefs about vaccines are the result of employing ineffective epistemic means to achieve their epistemic goals (e.g., trusting unreliable sources of information about vaccines and mistrusting trustworthy ones), not of pursuing of non-epistemic goals at the expense of epistemic ones.

---

[107] Counter-deniers might argue that, as long as her non-epistemic goals affect Helen's reasoning so that it is no longer appropriately responsive to her evidence, she is still a motivated reasoner. However, the literature on inductive risk convincingly questions the assumption that our non-epistemic goals should not affect how we respond to evidence (see, e.g., (Douglas 2009)). Moreover, as described, Helen's case gives us no reason to assume that Helen's non-epistemic goals impair her evaluation of the evidence. Counter-deniers might respond that the empirical literature on motivated reasoning suggests that people's non-epistemic goals impair their evaluation of the evidence. However, this seems to be a reason to believe that we are all motivated reasoners when we evaluate claims that are relevant to our practical decisions. So, if the charge of epistemic irresponsibility applies to misinformed non-believers like Helen, it also applies to most of us.

This argument seems to generalize to other prototypical high-stakes cases of (alleged) science denial, such HIV/AIDS denial, or certain forms of COVID scepticism. In those cases, too, it seems highly implausible to suggest that the alleged science deniers do not believe the relevant scientific claims due to motivated reasoning. Presumably, in high-stakes cases, people are trying their utmost to make decisions that are in their best interest (or that of their loved ones). Consider the case of Ivan, an HIV-positive patient who refuses antiretroviral drugs because he believes AIDS was invented by pharmaceutical companies to sell those drugs. Presumably, Ivan is not in denial about the truth of the claim that antiretroviral drugs are a safe and effective way to prevent or reverse AIDS, as that is not an inconvenient truth for an HIV-positive patient and refusing to believe it does not seem to serve any plausible non-epistemic goal of his. At most, one would expect Ivan to be in denial about there being no safe or effective treatment for his condition (if there wasn't one), not about there being such a treatment (given that there is one). A much more plausible diagnosis is that Ivan fails to believe in the safety and effectiveness of antiretroviral drugs because he has been exposed to misinformation on the topic of the kind spread by untrustworthy sources such as Joan Shenton's book *Positively False*.[108]

If, as argued in Section 1, the primary concern about science denial is that it is practically harmful, high-stakes cases are likely to raise the most serious practical concerns. This means that, if the arguments in this subsection are correct, the counter-deniers' practical concerns are largely misplaced. In cases in which someone's safety or well-being (or that of their loved ones) is at stake, misinformed non-belief seems a much more plausible diagnosis than motivated non-belief.[109]

Before concluding this subsection, let me mention two important points. The first point is that, while arguments in this subsection suggest that misinformed non-belief is a more plausible diagnosis than motivated non-belief in

---

[108] It might be objected that the argument does not generalize because people are willing to take risks with their own health that they are not willing to take with the health of their children, which means that my argument about Helen does not necessarily apply to Ivan. However, my main point is that one can fail to believe and act upon widely accepted scientific claims without being a motivated non-believer. The fact that some people are willing to take greater risks for themselves is no reason to believe that they are motivated non-believers – that is, that they engage in motivated reasoning when it comes to scientific claims that are relevant to practical decisions about their own well-being. At most, it seems to be a reason to believe that they are prudential non-compliers (see Section 26).

[109] I should note that the arguments in this subsection are compatible with the view that, once one becomes a science non-believer, motivated reasoning (or other cognitive factors (such as cognitive biases)) might contribute to their remaining a science non-believer. What I have claimed is that, in high-stakes cases, it seems very implausible to assume that the failure to accept the relevant scientific claims is the result of motivated reasoning instead of exposure to misinformation and that, therefore, misinformed non-belief is the more plausible diagnosis of the two.

high-stakes cases, this does not necessarily mean that it is the only or most common diagnosis in such cases. In many high-stakes cases, prudential non-compliance (Section 2.7) might be a better diagnosis. However, the argument is meant to show that either of these two diagnoses is more likely than a motivated non-belief diagnosis in many high-stakes cases. Moreover, even if misinformed non-belief and prudential non-compliance are distinct phenomena, there might be some similarities between them. In fact, in many cases, the difference between the two might be largely a matter of degree. For example, exposure to misinformation might lower someone's credence in a widely accepted scientific claim enough to turn them into a prudential non-complier but not enough to turn them into a misinformed non-believer. For another, exposure to misinformation might play an important in shaping the preferences of prudential non-compliers (e.g., affecting their perception of the severity of a disease).

The second and, in my view, most important point is that the discussion of misinformed non-belief suggests an alternative diagnosis of the collective epistemic malaise afflicting liberal democracies, which I call *the post-trust diagnosis*. Unlike the *post-truth* diagnosis, which claims that the malaise originates from a lack of interest in (or concern for) the truth among a growing number of our fellow citizens, the *post-trust* diagnosis maintains that it originates from a lack of trust in the institutions forming the backbone of the epistemic infrastructure of liberal democracies on the part of a growing number of communities. The *epistemic infrastructure* of a society is the system of institutions, norms, and practices that promote the reliable production, transmission, reception, and uptake of information and impede the production, transmission, reception, and uptake of misinformation. The institutions that form the backbone of the epistemic infrastructure of liberal democracies include not only science but also government and the press. According to the *post-truth* diagnosis, for example, vaccine hesitancy would be due to the fact that (allegedly) a growing number of people have no interest in (or concern for) the truth about the safety of vaccines. However, as argued in this subsection, this explanation seems to be much less plausible than the one offered by the *post-trust* diagnosis, which maintains that vaccine hesitancy is due to the fact that a growing number of communities do not trust science to conscientiously and objectively investigate the safety of vaccines, government to responsibly regulate their use, or the press to independently monitor those institutions and hold them accountable. This mistrust leaves those communities more vulnerable to misinformation about vaccines (as well as misinformation about those institutions).

3.4 Misinformed Non-Belief in Low-Stakes Cases

The previous subsection argued that, in high-stakes cases, misinformed non-belief and prudential non-compliance are much more plausible diagnoses than motivated non-belief. This subsection argues that misinformed non-belief might be a better diagnosis even in some low-stakes cases that, superficially, might appear to be particularly amenable to an explanation in terms of motivated reasoning One such case is the rejection of the theory of evolution by biblical literalists. Despite appearances, it is far from obvious that their disbelief stems from motivated reasoning aimed at protecting their religious beliefs. Presumably, some creationists are 'true believers', who do not engage in motivated reasoning at all. Consider, for example, the case of Karen, a woman who grew up in a Christian community that believes that the Bible is the word of an omniscient and benevolent God and that it is literally true in its entirety. Karen believes that the story of the creation as told by the Bible is the literally true account of the origins of life on Earth and wholly rejects the theory of evolution. Superficially, Karen's case may seem a prototypical case of motivated reasoning. It seems plausible to assume that, in reasoning about the origins of life, Karen aims to achieve the non-epistemic goals of protecting her religious beliefs or affirming her identity as a member of her religious community at the expense of the epistemic goal of having true beliefs about the topic. However, this does not need to be the case. In fact, for all we know, Karen's only goals in reasoning about the origins of life could be purely epistemic. Karen knows that most scientists disagree with her and, instead, believe a completely different explanation of the origins of life on Earth. However, the scientists themselves concede that science is fallible and that even their best theories could be false. God, on the other hand, is infallible, or so Karen believes. Given her background beliefs, it seems that, if Karen is exclusively interested in having true beliefs about the origins of life, she should believe the account found in the Bible, not the one widely accepted by scientists. While Karen does reject the theory of evolution because of her religious beliefs, this need not be due to motivated reasoning. In fact, we can assume that her goals are purely epistemic. Karen believes that she has a better chance of achieving her purely epistemic goals by believing the truthful word of an infallible God rather than the defeasible theories of fallible scientists. Once again, if Karen's conduct is epistemically irresponsible (or blameworthy), it is not because she engages in motivated reasoning, but because she too, like Frank (Section 3.2) and Helen (Section 3.3), trusts sources, such as the Bible and her minister, that are not trustworthy about this specific topic and she mistrusts sources that are more trustworthy on this topic, such as the vast majority of biologists.

Of course, this is merely a fictional example. It is possible that most or all real-world creationists are motivated non-believers rather than misinformed non-believers like Karen. However, Karen's case still teaches us two important lessons. First, it suggests that, even if all actual creationists are motivated non-believers, they are likely to be 'impure' motivated non-believers, as it is undeniable that creationists are typically exposed to large amounts of misinformation about specific scientific topics (as well as about higher-order misinformation about the nature of scientific inquiry and its aims) and implausible to assume that this misinformation has no effect on their beliefs. Second, it shows that even cases that, at first, appear particularly amenable to explanations in terms of motivated reasoning may, on closer scrutiny, turn out to be cases of misinformed non-belief (or, at the very least, cases of 'impure' motivated non-belief). In other words, the mere failure to believe an inconvenient truth is not sufficient to establish a motivated non-belief diagnosis. More evidence is needed to avoid misdiagnosing cases of misinformed non-belief (or 'impure' motivated non-belief) as 'pure' cases of motivated non-belief. In the next subsection, I argue that this evidence is likely to be particularly elusive in low-stakes cases.

3.5 Is Motivated Non-Belief Discernible from Insincere Contestation?

In the previous subsection, I have argued that, in order to reliably discern misinformed non-believers from motivated non-believers in low-stakes cases, more evidence is needed that their beliefs are actually the result of motivated reasoning. In this subsection, I argue that this evidence is likely to be particularly elusive in those cases. In low-stakes cases, it is extremely difficult to distinguish alleged cases of motivated non-belief from cases of insincere contestation. If my arguments are correct, then motivated non-belief diagnoses in low-stakes cases are threatened on both flanks. On one side, as we have seen in the previous subsection, ('pure') motivated non-belief might be difficult to discern from misinformed non-belief (or 'impure' motivated non-belief). On the other side, it is difficult to discern from insincere contestation.

To see the last point, consider the cases of Ben and Jerry. Ben is a motivated non-believer. He is libertarian who works as an engineer for an oil company, drives a gas-guzzling pickup truck, and supports political candidates who share his views about free markets and small government. Despite his engineering background and the fact that he knows more about climate science than most, Ben is in denial about the reality of anthropogenic climate change because his reasoning about it prioritizes the non-epistemic goals of protecting his political beliefs and of justifying his career and lifestyle over the epistemic goal of having true beliefs about the climate. As a result, Ben sincerely believes that the evidence for anthropogenic climate change is weak and that the climate scientists who claim otherwise do so either as a matter of professional interest (e.g., to receive large grants) or to promote their own political agendas (e.g., they are environmentalists or socialists, who support Big Government and oppose free markets). However, ironically, it is, in fact, him who is trying to protect his economic interests and his political beliefs through motivated reasoning. Jerry, on the other hand, is exactly like Ben except for the fact that, unlike Ben, he is an insincere contester. Jerry does not engage in motivated reasoning on climate change. In fact, he is privately inclined to believe in the reality of anthropogenic climate change. However, he claims not to do so to further his own political and economic interests.

The problem for counter-deniers is that it might be extremely difficult to tell Ben and Jerry apart. In other words, it is extremely difficult to determine whether someone like Ben genuinely does not believe in anthropogenic climate change or (like Jerry) merely claims not to believe in it to serve his own non-epistemic goals (e.g., justify his lifestyle decisions, his opposition to policies with which he disagrees on political or ideological grounds, or even to just protect his well-paid job). For all we know, people who behave like Ben and Jerry might be insincere contesters (like Jerry) rather than motivated non-believers (like Ben).

The problem here is partly that the verbal behaviour of Jerry is likely to be indistinguishable from that of Ben. Talk is cheap. If Jerry were asked whether he really does not believe in anthropogenic climate change, he would insist that he *really* does not accept the reality of anthropogenic climate change. However, in light of Jerry's own claims, Jerry's denials would ring hollow. After all, Jerry (like Ben) would be the first to suggest climate scientists say things that they do not actually believe to serve their own political or professional interests, so why would the same not apply to Jerry?

This worry seems to be particularly troublesome given that one of the main ways to estimate the prevalence of science denial is to conduct surveys (such as the ones run periodically by the Pew Research Center in the United States).[110] The problem is that, if, as counter-deniers seem to believe, one of the primary goals of motivated non-believers is to affirm their social identity, then it seems plausible that someone like Jerry might affirm his social identity by not responding truthfully to surveys and claiming not to believe in anthropogenic climate change even if he does. This is particularly true if he sees this as

---

[110] See, e.g., (Tyson, Funk, and Kennedy 2023).

a chance to provoke the sort of 'soy-latte-sipping liberals' who, in his mind, read these surveys and fret about the rise of climate-change denial. It is likely that most extant surveys miscount insincere contesters as science deniers and, as a result, grossly overestimate the prevalence of science denial (understood as motivated non-belief).[111]

Moreover, in low-stakes cases, the non-verbal behaviours of alleged science deniers are also likely to be virtually indiscernible from those of insincere contesters. To see why, compare Jerry's case to a high-stakes case – that of Logan, a tobacco executive who quit smoking despite publicly downplaying the health risks of tobacco. While Logan's verbal behaviour might suggest that he is a science non-believer, his non-verbal behaviour is a tell-tale sign of the insincerity of his science contestations. If Logan actually believed his own claims, he would put his lungs where his mouth is. In low-stakes cases (like those of Ben and Jerry), on the other hand, it is very unlikely for there to be any non-verbal tell-tale sign. Most of Ben's and Jerry's choices (e.g., what vehicle they drive, what job they have, what political candidates they support) are likely to be the result of a wide variety of factors both within and outside their control and their beliefs about anthropogenic climate change are likely to play a relatively minor role in those decisions and, at any rate, they are very unlikely to be a decisive factor. In other words, it seems implausible to assume that, if Ben were to believe in anthropogenic climate change (despite his science contestations), he would drive a different vehicle than Jerry, have a different job than Jerry, or support different political candidates than Jerry.

These considerations suggest that, despite their different beliefs about climate change, Ben's and Jerry's behaviours (both verbal and non-verbal) might be completely indistinguishable from one another. There is no tell-tale sign of the insincerity of Jerry's science contestation or of the sincerity of Ben's disbelief. In fact, only Jerry can know that he is an insincere contester. Ben, on the other hand, does not even know he is a motivated non-believer, because, as long as he is successfully in denial about the reality of anthropogenic climate change, he must also be in denial about being in denial.[112]

---

[111] On the influence of expressive responding as well as approaches to mitigating its influence, see, e.g., (Litman et al. 2023).

[112] Counter-deniers might complain that the arguments in this subsection are likely to overgeneralize wildly. Can we ever know for sure what other people actually believe? However, the scope of the arguments in this subsection is restricted in at least two respects. First, it only applies to beliefs that are not directly connected to the subject's practical decisions in high-stakes or medium-stakes cases. Second, it only applies to beliefs for which we have reason to suspect the speakers may not be sincere (as indicated, for example, by the literature on expressive responding (see, e.g., (Litman et al. 2023))).

3.6 Who Needs to Be a Motivated Non-Believer?

This subsection argues that someone like Ben does not need to engage in motivated reasoning to pursue the non-epistemic goals that are usually attributed to motivated non-believers (and other motivated reasoners). Those goals could be pursued equally well (and, possibly, more cheaply) in other ways. Here, I discuss the two sorts of non-epistemic goals that are most often attributed to motivated non-believers. The first is to affirm their social or political identities. The second is to protect their social, political, or religious beliefs. Note that the arguments in this subsection only target those who, like counter-deniers, adopt a particularist conception of motivated reasoning and apply only to people who are (supposedly) 'pure' motivated non-believers (i.e., they are motivated non-believers who have not been exposed to misinformation on the relevant scientific topics).

Let me consider first the goal of affirming one's social or political identity. We are social animals and, as such, we use all kinds of means (e.g., items of clothing, flags, words, actions) to signal our belonging to certain social groups (e.g., Republicans or Democrats, rural or urban, working-class or middle-class, Evangelical Christian or Hindu, Lakers fan or Celtics fan). According to counter-deniers, some of the science deniers' beliefs are signals of this kind. For example, Matt might reject anthropogenic climate change in order to affirm his socio-political identity as a rural working-class conservative man. 'People like him' don't believe in anthropogenic climate change! However, Matt does not actually need to actively disbelieve the reality of anthropogenic climate change in order to successfully signal his belonging to a socio-political group. He can do so equally well by faking the signal – that is, by merely claiming to disbelieve the relevant scientific claims. In fact, the latter option would seem cognitively cheaper. I take it that belief is largely involuntary and, since we are assuming that Matt is a 'pure' motivated non-believer, he has not been exposed to any misinformation about climate change.[113] This means that either Matt has not been exposed to any information about it (and thus has no opinion on it), or he has only been exposed to truthful information about it (and thus he is inclined to believe in anthropogenic climate change). In either case, his disbelieving the reality of anthropogenic climate change would require an effort to believe against the evidence (or lack thereof). This is why Matt would supposedly have to engage in motivated

---

[113] Note that this is a thought experiment, not a realistic example. In reality, it seems impossible for someone like Matt not to be at all exposed to misinformation about climate change. After all, if Matt is aware of the fact that 'people like him' do not believe in anthropogenic climate change, it must be because people who belong to the social groups with which Matt identifies publicly reject anthropogenic climate change, which, in itself, is enough for Matt to be exposed to misinformation about climate change.

reasoning by, for example, seeking out misinformation that to support disbelief in anthropogenic climate change. However, that requires time and effort. Matt, however, does not need to go through the effort required to believe against the evidence. He could achieve the goal of affirming his social identity equally well (and cognitively more cheaply) by being an insincere contester.

Counter-deniers might reply that self-deception is cognitively cheaper than deception and that people typically deceive themselves in order to better deceive others.[114] However, for one thing, it is unclear that is generally true, and, for another, even if it is, there is no reason to think that Matt must engage in deception to be an insincere contester. Matt could fake the signal equally well (and more cheaply) by bullshitting about anthropogenic climate change rather than lying about it.[115] This diagnosis seems to fit particularly well those cases in which people switch between different claims ('climate change is not happening', 'it is happening but it is not caused by human activity', 'it is caused by human activity, but it is no cause for alarm', etc.).

Consider now the goal of protecting one's social, political, or religious beliefs. Again, it seems that this non-epistemic goal can be served equally well (and cognitively more cheaply) by other means. Suppose that Ben has not been exposed to any misinformation about climate change. Like Matt, he would have to disbelieve the reality of anthropogenic climate change despite the evidence, and like in Matt's case, this would require him to seek out misinformation about climate change, which requires time and effort. However, no such effort would be needed to square Ben's political beliefs in free markets and small government with the reality of anthropogenic climate change, as there are plenty of other ways for him to do so. For example, Ben could simply believe that the best response to anthropogenic climate change is to adapt to it rather than trying to mitigate it.

To my mind, the best explanation of why, in the real world, people like Ben tend to take the more arduous route of disputing the reality of anthropogenic climate change despite the scientific consensus about it is that, unlike Ben, they have been exposed to misinformation about it from a variety of sources they trust. The division of epistemic labour applies also to questions of policy. The evidence seems to suggest that most people defer to their favourite politicians, political commentators, or political parties when it comes to deciding what political issues are important and which policies would best address those issues.[116] This

---

[114] They could support this claim appealing to Robert Trivers' work on self-deception (see, e.g., (Trivers 2011)).

[115] Here, I am employing (a broader version of) the standard notion of bullshit as introduced by Henry Frankfurt (2005).

[116] See, e.g., (Achen and Bartels 2016, chap. 10).

Science Denial 51

suggests that, if those who have a vested interest in delaying the adoption of policies to mitigate climate change had adopted early on the strategy of advocating for adapting to climate change (instead of the strategy of disputing the scientific status of anthropogenic climate change), it is highly unlikely that people like Ben would have disbelieved the reality of anthropogenic climate change instead of believing in it but disputing that mitigation is the best response to it.[117]

3.7 Is Motivated Non-Belief Practically Harmful?

Earlier (in Section 3.3), I argued that misinformed non-belief is a more plausible diagnosis than motivated non-belief in high-stakes cases of science non-belief and also that this shows that counter-deniers' practical concerns are largely misplaced, as the high-stakes cases are the ones that are more likely to have practically harmful consequences. Counter-deniers, however, might argue that motivated non-belief can still be practically harmful in low-stakes cases even if the practical harms in these cases are likely to be more indirect and diffuse than in high-stakes cases.[118] For example, they might argue that, as a result of his motivated non-belief, Ben is more likely to make practically harmful choices in his capacity as a consumer (e.g., buying a fuel-inefficient car, travelling more by plane) or as a citizen (e.g., support political candidates who oppose climate mitigation policies). While the practical harms caused by Ben's choices might be negligible in the grand scheme of things, they still contribute to harming our planet and everyone on it, including Ben himself.

The main problem with this line of reasoning is that it presupposes that, in cases of motivated reasoning, the relationship between beliefs and actions is the same as in ordinary cases. However, this assumption does not seem to withstand scrutiny. If and when people engage in motivated reasoning, they seem to do so in an attempt to rationalize their pre-existing beliefs, preferences, and decisions.[119] Consider again the case of Andy. Presumably, the reason why Andy is in denial about his drinking habits is that, if he wasn't, then, at least, he would have to acknowledge the fact that his drinking habits are negatively affecting him and those around him in various ways, which might make him feel guilty and

---

[117] While Daniel Williams (2023b) might be right in thinking that, on many topics, people act as consumers in markets for rationalizations, in my view, he is wrong in thinking that rationalization markets are driven primarily by demand rather than supply. For one thing, in liberal democracies, rationalization markets tend to be oligopolies that, as the case of misinformation about climate change shows, produce rationalizations for beliefs convenient to the producers (and their political and economic backers) rather than beliefs convenient to their consumers (see (Contessa 2022)).

[118] For a variety of views of our individual responsibilities with regard to climate change, see, e.g., (Sinnott-Armstrong 2005), (Hiller 2011), and (Almassi 2012).

[119] Here, I understand rationalization broadly enough so as to include those cases in which the rationalization of pre-existing beliefs, preferences, and decisions strengthens those beliefs, preferences, or decisions. While this weakens my argument somewhat, it does not undermine it.

embarrassed. Moreover, he might have to do something about it, which he might not be willing or able to do, which might make him feel inadequate and powerless. Being in denial allows Andy to keep doing what he does while protecting him from the negative feelings that would be likely to arise if he acknowledged his drinking problem and justify his inaction to himself and to others. In other words, Andy's motivated reasoning is primarily a source of post hoc rationalizations (i.e., rationalizations of his pre-existing beliefs, preferences, and decisions). But this seems to suggest that, when applied to motivated non-belief, the counterdeniers' practical concern stems from a misconception of the relationship between beliefs and decisions in motivated reasoning.

Our practical decisions as consumers and as citizens are typically the result of a wide variety of factors both within and outside our control and, especially in low-stakes cases, it seems very unlikely that our beliefs about scientific claims play a significant role in those decisions and even more unlikely that they play a decisive role – that is, that if we were able to change the minds of motivated non-believers, they would be making different practical decisions. Consider, for example, the case of air travel. Anecdotally, the vast majority of the academics I know (myself included) believe in anthropogenic climate change, and yet, even if it is well-known that air travel is one of the main contributions individuals can make to global carbon emissions, most academics I know (including, admittedly, myself) do not drastically limit their work-related air travel. Systematic empirical studies seem to support my observations (for both academics and non-academics).[120] In fact, a recent study found that climate scientists fly more for work than other academics.[121]

This in itself is a major problem for the argument we are considering, as it suggests that, even if Ben is indeed a motivated non-believer, his behaviour would not change significantly if only he started believing in anthropogenic climate change. In particular, it is unlikely that, even if we could magically change his mind about anthropogenic climate change, he would buy a different car, get a different job, or support different political candidates. Ultimately, whether or not, deep down, Ben believes in anthropogenic climate change seems to be largely irrelevant to his behaviour as a citizen or as a consumer,

---

[120] For academics, see, e.g., (Kreil 2021), (Thaller, Schreuer, and Posch 2021), and (Tseng, Lee, and Higham 2022); for non-academics, see, e.g., (Hares, Dickinson, and Wilkes 2010). Note that this does not apply to all decisions as consumers. For example, people who believe in anthropogenic climate change are more likely to drive an electric vehicle than people who do not. However, this still does not mean that their belief in climate change is a decisive factor in their decision to buy an electric vehicle (as opposed to, e.g., the desire to signal one's belonging to certain social groups). Air travel is a particularly interesting test case because, as of yet, little or no social stigma seems to be attached to it even among people who believe in anthropogenic climate change.

[121] (Whitmarsh et al. 2020).

as, in low-stakes cases, there are plenty of other potential rationalizations Ben's could employ to justify (or excuse) his pre-existing practically harmful decisions to himself or to other.

It is important to note that the arguments in this subsection are not meant to establish that motivated non-belief is completely practically harmless. For one thing, it is plausible to suppose that motivated reasoning strengthens a person's confidence in their own beliefs, preferences, and decisions. Their upshot, in my view, is that, even if motivated reasoning has these effects, there is still little or no reason to believe that it ever plays a *decisive* role in practical decisions in low-stakes cases, as, arguably, these are affected by a wide variety of much weightier factors, such as social and economic factors. If so, then, in low-stakes cases, motivated non-belief is, at most, a fig leaf to cover up one's true motivations. If this is correct, then the counter-deniers' practical concerns would be misplaced even in genuine cases of motivated non-belief.

3.8 What to Do about Motivated Non-Belief?

Even if my arguments so far are wrong and motivated non-belief makes a decisive contribution to practically harmful decisions, counter-deniers still face a dilemma. On the one hand, insofar as motivated non-believers' decisions are practically harmful to others, it is unclear that we should solely try to persuade them to make different decisions. After all, we do not merely try to persuade smokers not to expose others to the practically harmful consequences of their smoking – we forbid them to smoke in public places. One of the central tenets of liberalism is that people should be free to make their own decisions but only as long as those decisions do not harm others. On the other hand, if motivated non-believers' decisions are exclusively harmful to them, then it is unclear that there is anything we should do about it (other than, possibly, trying to convince them not to make those decisions by means of rational persuasion).[122] Another central tenet of liberalism is its anti-paternalism – competent adults should be free to make their own decisions even if those decisions might have practically harmful consequences for them (as long as they don't harm anyone else). For example, even if smoking does have practically harmful consequences for smokers, we do not forbid smokers from smoking in their own homes. At most, we try to persuade them to quit smoking.

Arguably, the best response to this dilemma is to try to blunt its horns. The first part of this response aims to blunt its first horn. While it is possible to limit the freedom of people when their actions might harm others, there must be some

---

[122] Not everyone, of course, would agree with this. For two alternative views on rational persuasion and its legitimacy, see (Tsai 2014) and (McKenna 2020).

proportionality between the harm those decisions would cause to others and the harm caused to them by curtailing their freedom to make those decisions. For example, even if the seasonal flu kills people every year, we collectively judge that the harms of limiting the freedom of flu patients by quarantining them would outweigh the harms caused to the rest of society. This, however, does not mean that we should not try to encourage flu patients to take steps to avoid spreading the flu. In other words, the short reply to the first horn of the dilemma is that, in many cases, there is room for non-coercive interventions. The second part is that, while liberalism opposes paternalism in theory, liberal democracies often adopt paternalistic policies in practice. The standard example of such policies are laws enforcing the use of seatbelts. Again, the issue seems to be one of proportionality. We collectively judge that the practical harms caused by letting people freely decide whether to wear a seatbelt far outweigh the harms caused by the minimal invasion of freedom that is caused by imposing the use of seatbelts.

While this response blunts the horns of the dilemma, it does so at the cost of raising the bar for counter-deniers. Counter-deniers must not only show that motivated non-belief is practically harmful (despite the arguments to the contrary discussed earlier) but also that the practical harms it supposedly causes fall within the grey area between what can be legitimately regulated by a liberal democracy and what cannot. However, once we look more closely at cases of (alleged) motivated non-belief, it is unclear that they typically fall within grey area. Suppose, for example, that, contrary to my arguments, Ben's motivated non-belief is a decisive factor in his practically harmful decisions as a consumer (e.g., which vehicle to purchase) or as a citizens (e.g., which political candidate to support). Neither kind of decisions seems to fall into the grey area between what can be legitimately regulated by a liberal democratic government and what cannot. On the one hand, Ben's decision to buy a highly fuel-inefficient vehicle is practically harmful not only to him but also to everyone else and, in fact, to the whole planet. It is, therefore, a clear instance of the kind of decision that can be legitimately restricted in a liberal democracy (by, e.g., increasing fuel-efficiency standards). On the other hand, his decision to support a political candidate is an instance of the kind of decision that a liberal democracy cannot legitimately restrict (except in extreme cases), as a political system that does not allow people to support their preferred political candidate is not truly a democracy, let alone a liberal one.[123]

3.9 Conclusion

In this section, I have argued that, despite the promise of the motivational account, counter-deniers face a dilemma. On the one hand, while motivated

---

[123] Possible exceptions to this general principle are openly anti-democratic candidates.

non-belief might be always epistemically irresponsible (or blameworthy), it is dubious that it is ever practically harmful. On the other hand, while misinformed non-belief may sometimes be practically harmful, it is doubtful that it is always (or ever) epistemically irresponsible (or blameworthy).

More specifically, I have argued:

- that misinformed non-belief is a much more plausible diagnosis than motivated non-belief in high-stakes cases, such as vaccine hesitancy or HIV/AIDS denial,
- that misinformed science non-belief might be the correct diagnoses even in low-stakes cases that superficially appear particularly amenable to an explanation in terms of motivated reasoning, such as creationism,
- that, in low-stakes cases, motivated non-belief might often be indiscernible from insincere contestation, which, among other things, makes it difficult to determine the actual prevalence of motivated non-belief,
- that there are cognitively cheaper ways to achieve the non-epistemic goals usually attributed to motivated non-believers,
- that it is unclear that, in low-stakes cases, motivated non-belief is as practically harmful as counter-deniers assume, as it might be just a way for the alleged motivated non-believers to rationalize their own pre-existing beliefs, preferences, and decisions rather than a decisive factor in those decisions, and
- that, even if motivated non-belief is, in fact, practically harmful, its practical harms are either of the kind that may be prevented through the legitimate use of coercion or of the kind that cannot be prevented without undermining the democratic nature of the political system.

While much more could be said about each of these theses, I think that taken together they form a strong presumptive case against the view that science denial (understood as motivated non-belief) warrants both the epistemic concerns and the practical concerns of counter-deniers. I take it that there are a few lessons to be learned from this discussion.

The first lesson is that, if science denial is identified with motivated non-belief, then it becomes exceedingly difficult to distinguish it from a number of superficially similar phenomena, such as misinformed non-belief, insincere contestation, or prudential non-compliance. The narrative that science denial is on the rise is supported primarily by three types of empirical evidence: anecdotal evidence, direct systematic evidence (e.g., opinion surveys), and indirect systematic evidence (e.g., evidence about vaccination rates). However, once those specific phenomena are distinguished from one another, it becomes apparent these bodies of evidence do not give us a reliable way to differentiate among them. People seem to have all kinds of reasons to engage in science contestation in certain conversational contexts (such as opinion surveys) and, even more so, to act as they do in a variety of practical contexts (such as when making decisions about vaccinations). I think the lesson here is that we should try to better understand the beliefs and concerns of each alleged science denier by actually engaging with them (rather than reproaching them from afar as counter-deniers are wont to do).

The second lesson is that, if science denial is identified with motivated non-belief, then it is doubtful that it is ever practically harmful, as it is unlikely to be a decisive factor in the practically harmful decisions of the (alleged) motivated non-believer. It is much more likely that motivate reasoning helps them produce post hoc rationalizations of their practical decisions.

The third lesson is that, unlike motivated non-belief, misinformed non-belief and prudential non-compliance are likely to be practically harmful in high-stakes cases (such as those of childhood vaccinations, COVID-19, or HIV/AIDS). However, neither is plausibly construed as a form of science denial. For one thing, it is doubtful that either is always epistemically irresponsible or blameworthy, and, even when they are, the epistemic irresponsibility involved is different both in degree and in kind from the kind of epistemic irresponsibility allegedly involved in motivated non-belief. For another, insofar as motivated non-believers are epistemically irresponsible, it seems to be because they place more epistemic trust in epistemically unreliable sources than in epistemically trustworthy ones. What I think we should learn from this is that we need much finer-grained categories than 'science denial' to be able to properly identify the sources of potentially harmful behaviour in high-stakes cases and develop effective approaches to addressing them appropriately and respectfully. For example, we need a better appreciation of the fact that vaccine hesitancy is a complex and multifaceted phenomenon and that to brand all vaccine hesitators as science deniers is likely not only inaccurate but also counterproductive.[124] A much more promising approach would be to start from trying to understand vaccine hesitators' concerns in an empathetic and non-judgmental manner and then trying to address them.[125]

The fourth and final lesson is that science contestation (whether sincere or insincere) is likely to be epistemically harmful, as it pollutes the epistemic environment with the kind of misinformation that seem to feed misinformed

---

[124] See (Goldenberg 2021) for an enlightening and thorough philosophical discussion of the phenomenon in all its complexity. See (Holford et al. 2023) and (Fasce et al. 2023) for an empirical approach to developing a taxonomy of sources of vaccine-hesitancy.

[125] The Empathetic Refutational Approach (which employs two-way face-to-face conversations with vaccine hesitators in which their specific concerns are elicited, acknowledged, and then addressed) seems to be a promising example of this approach to misinformed non-belief and science under-belief (see, e.g., (Holford et al. 2024)).

non-belief and certain forms of prudential non-compliance. In fact, if we do not want to abandon the pejorative term 'science denial', science contestation (especially in its insincere form) is arguably the phenomenon that is most deserving of it. However, if the post-trust diagnosis is correct, then the most promising approach to preventing science contestation from feeding misinformed non-belief and other practically harmful behaviours is not to try to prevent individuals from being exposed to misinformation (which, in a liberal democracy, is neither feasible nor desirable) but to immunize communities from misinformation by restoring their trust in the institutions that underpin the epistemic infrastructure of liberal democracies. As I mention in the Conclusion, however, this is easier said than done, as their mistrust is not always entirely unjustified or misplaced.

Conclusion: Beyond 'Science Denial'

In this Element, I have argued that the notion of science denial is highly ambiguous and that, once we try to disambiguate it, it becomes clear that a wide variety of distinct phenomena are often conflated under that label. If we want to better understand (and, if needed, address) each of these phenomena, we need to carefully distinguish them from one another and not lump them all together under the same label. Once we carefully distinguish among these various phenomena, the one that seems to be the best candidate for the role of science denial – namely, motivated non-belief – does not appear to warrant the practical concerns that animate counter-deniers. The phenomena that most seems to warrant those concerns – namely, misinformed non-belief and prudential non-compliance – cannot be construed as forms of science denial, as they do not raise the same kinds of epistemic concerns as motivated non-belief. To my mind, these arguments give us solid reasons to abandon the term 'science denial' altogether.

This conclusion is strengthened by worries about the epistemically and practically harmful consequences of using and misusing terms such as 'science denier'. Let me mention three of them. First, as noted earlier (Section 2.1), charges of science denial are often used to either depoliticize political questions or improperly politicize scientific ones. Both abuses can have epistemically and practically harmful consequences, as they can negatively affect scientific inquiry, public policy debates, and policymaking.

Second, due to their negative connotations, terms such as 'science denial' are unlikely to serve the counter-deniers stated goals. If the ultimate goal of counter-deniers is to persuade people to get themselves or their children vaccinated or to impress on them the urgency of enacting policies to mitigate anthropogenic climate change, then it seems unlikely that branding them as science deniers is an effective means to achieve those goals. In fact, it is likely to be counterproductive, as the use of pejorative terms is likely to reduce the chances of persuading them to make different personal choices and to undermine efforts to find common ground for agreement over the relevant issues.

Third, pejorative terms are likely to exacerbate affective polarization.[126] For example, during the COVID-19 pandemic, the deaths of COVID sceptics were often treated insensitively and inhumanely, including cases where their deaths were openly ridiculed or the families of the deceased were subjected to online harassment.[127] A recent study found that people who identify as Democrats or who are vaccinated are more likely to perceive the deaths of people who refused COVID vaccines as justified.[128] This seems to show that labels such as 'anti-vaxxer' or 'science denier' can have polarizing and dehumanizing effects, which is concerning in itself.[129] Moreover, increased level of affective polarization might have the unwanted effect of increasing both sides' susceptibility to misinformation.[130]

In light of these considerations, my recommendation would be to abandon the term 'science denier' and its ilk altogether.[131] However, I doubt that this is likely to happen in the short-term. These labels seem to be too deeply ingrained in public discourse, and, admittedly, they are catchier than any of the labels I have used to identify more specific phenomena. As far as I can see, the second-best option is to use 'science denial' to refer to (presumptively) insincere forms of science contestation – particularly, those that appear to be coordinated and/or motivated by social, economic, or political interests.[132] The now-standard example of this kind of insincere science contestation is the disinformation campaign orchestrated by the tobacco industry to deceive and confuse the

---

[126] On the increase of affective polarization in the United States, see, e.g., (Iyengar and Westwood 2015).
[127] See, e.g., (Rascoe 2020) and (Levin 2021). [128] (Grizzard, Frazer, and Monge 2023).
[129] On the pernicious effects of dehumanization, see, e.g., (Smith 2021b).
[130] See, e.g., (Jenke 2024).
[131] I take my position with regard to 'science denier' to be analogous to the one advocated by Joshua Habgood-Coote (2019) with regard to 'fake news' and 'post-truth'. I also see my arguments here to complement his arguments against the post-truth diagnosis (as well as those of Michael Hannon (2023)). As long as 'post-truth' has any descriptive content, it seems to refer to the general phenomenon of which motivated non-belief seems to be a species.
[132] Some authors seem to already use the label this way (see, e.g., (Hoofnagle and Hoofnagle 2007), (Mann and Toles 2016, 53), and (Stott 2021, 6)). In fact, even counter-deniers sometimes use it this way (see, e.g., (McIntyre 2019, 155)).

general public about the health risks of second-hand smoking.[133] Insincere contestation is not only morally and epistemically blameworthy but also likely epistemically harmful, as it pollutes the epistemic environment with disinformation that likely feeds practically harmful forms of misinformed non-belief and prudential non-compliance. In its case, the use of labels with a strongly negative evaluative content seems more than amply justified. However, in order to clearly differentiate this specific use of the label from its other uses, it would be advisable to use it as an adjective rather than a noun (e.g., 'science-denial campaigns' or 'the science-denial industry').

Whatever the fate of terms such as 'science denial', the arguments in this Element strongly suggest that the worst option is to use 'science denial' either as a generic term to cover a variety of distinct phenomena or as an ambiguous term that may refer to different phenomena in different contexts.[134] To my mind, this is also the strongest argument against any attempt to salvage a purely descriptive use of the term (i.e., one devoid of any negatively valenced evaluative or expressive content). Even if this kind of reclamation project were feasible despite the strong connotations the term currently has, it is unlikely that it would be fruitful, as one of the main problems with it concern its lack of a clear descriptive content. It is unlikely that such a term can be useful in the investigation of the relevant phenomena. As I have argued, in order to understand (and, if necessary, address) phenomena such as motivated non-belief, misinformed non-belief, or prudential non-compliance, we need to carefully distinguish them from one another. A purely descriptive use of 'science denial' would still be likely to encourage the conflation of distinct phenomena and conceal the all-too-important differences among them. If anything, it is likely that we need a finer-grained taxonomy than the one offered here.

However, to my mind, one of the most important lessons to draw from this discussion is that counter-deniers presuppose an overly simplistic picture of the relationship between science and society.[135] This is the kind of picture also held by those who say things like 'science tells us' that something or other is the case or who exhort us to just 'follow the science'. I like to call the underlying picture *the oracular view of science* because it treats science as akin to *an oracle* – an epistemic authority whose pronouncements we have pre-emptive reasons to

---

[133] See, e.g., (Michaels 2008, chap. 7) and (Oreskes and Conway 2010, chap. 5). See (Kourany and Carrier 2020) for an edited volume that collects a number of philosophical contributions focusing on these issues.

[134] See, e.g., (McIntyre 2019, 149–51).

[135] For an overview of the current literature on the relationship between science and the public, see (Potochnik 2024) in this series.

believe.[136] I think that we have excellent reasons to reject the oracular view. Let me briefly mention five here. First, science does not 'tell' us anything because science itself cannot speak. Insofar as science 'speaks', it does so through its unofficial representatives, such as individual scientists or scientific organizations. Second, insofar as those unofficial representatives do speak for science, they don't always speak in a unified 'voice'. Disagreements are ubiquitous in science, and, arguably, they crucially contribute to its ability to self-correct. Third, even when its representatives do speak in a relatively unified 'voice', they can still get it wrong. Scientific inquiry is valuable not because it is an infallible source of immutable knowledge, but because, if properly practiced, it is a reliable and self-correcting source of actionable knowledge. Fourth, even when science's unofficial representatives do 'speak' in a relatively unified 'voice' about largely settled topics, science does not 'tell' us what we ought to do[137] or think.[138] Finally, ordinary people are unlikely to be able to reliably determine who, among those who claim to speak for science on any given scientific topic, should be trusted to do so responsibly.[139]

To my mind, this last point is particularly important. As I have argued, one of the phenomena that are most likely to be practically harmful, misinformed non-belief, is often the result of misplaced trust. Misinformed non-believers fail to believe widely accepted scientific claims as a result of exposure to misinformation from sources they trust. However, as I have also argued, it would be wrong to fault them for misplacing their trust. All on our own, we are incapable of discerning genuine information from misinformation or trustworthy sources from untrustworthy ones. Our only realistic option is to rely on our communities

---

[136] A pre-emptive reason to φ is a reason to φ that displaces all other reasons one might have to (or not to) φ (see, e.g., (Raz 1985, 10)). It is important to notice that this conception of epistemic authority is stronger than the already very strong one defended by Linda Zagzebski, who maintains that we have pre-emptive reasons to believe whatever an epistemic authority believes (not whatever an epistemic authority asserts) (see, in particular, (Zagzebski 2012)).

[137] The ubiquitous slogan 'follow the science' seems to imply that science can tell us what to do. However, even assuming the science is an epistemic authority, it is not a practical authority, and it is usually (and, in my view, correctly) assumed that the role of (legitimate) practical authorities (not that of epistemic authorities) to tell us what to do. On the distinction between practical and epistemic authorities, see (Raz 1986) (who, however, uses the label 'theoretical' instead of 'epistemic').

[138] As far as I know, no one explicitly defends the conception of epistemic authority that presupposed by the oracular view. Even Linda Zagzebski, who holds a superficially similar view, assumes that we have pre-emptive reasons to believe whatever epistemic authorities believe (as opposed to what they say) (see, in particular, (Zagzebski 2012)). For alternative views, see, e.g., (Lackey 2018) and (Croce 2018).

[139] Philosophers often tend to underestimate both the enormity and the difficulty of this task (see, e.g., (Anderson 2011)). As I argued elsewhere (Contessa 2023), it seems unreasonable to expect ordinary citizens to have the resources or the motivation to reliably form accurate assessments of the trustworthiness of a variety of sources that are alleged experts on the relevant topics.

to monitor the trustworthiness of various sources on our behalf. Misinformed non-believers happen to have the misfortune of inhabiting communities that are not epistemically discerning (i.e., communities in which the reputation of various sources does not reliably track their actual trustworthiness). While this lack of epistemic discernment can sometimes be due to the community being manipulated or deceived by bad actors with nefarious purposes, this is not always the case. Crucially, misinformation can only find fertile ground in communities that already mistrust the institutions that form the backbone of their society's epistemic infrastructure. As argued earlier (Section 3.3), for example, misinformation about vaccines is unlikely to take root in communities that do not already mistrust science to investigate the safety of vaccines, government to regulate their use, and the press to monitor those institutions and hold them accountable. Unlike the *post-truth* diagnosis, which claims that our collective epistemic malaise is due to a lack of interest in (or concern for) the truth (or other epistemic goals) on the part of many of our fellow citizens, the *post-trust* diagnosis argues that it is primarily due to a lack of trust in the institutions that are instrumental to the functioning of the epistemic infrastructure of liberal democracies.

What's more, according to the post-trust diagnosis, this mistrust is not always entirely unjustified or misplaced. Let me briefly mention three ways in which this is so in the case of science. First, historically, science and scientists have contributed – and, to some extent, still contribute – to the stigmatization, discrimination, marginalization, and abuse of historically oppressed communities, including women, racial and ethnic minorities, people with disabilities, and LGBTQ+ individuals.[140] Understandably, this has fostered mistrust towards science and scientists in those communities. Second, when non-epistemic values influence scientific decisions (from research priorities to the acceptance or rejection of scientific hypotheses),[141] communities that are underrepresented and/or marginalized among scientists may reasonably question whether the values employed in making those decisions align with their own. Third, when scientific research is conducted under (real or apparent) conflicts of interest,[142] communities might reasonably doubt the credibility of that research.

The post-trust diagnosis suggests that, to improve the uptake of scientific information by society, we should stop focusing on science non-belief, as an individual-level phenomenon and, instead, turn to public trust in science as a social-level phenomenon. On this view, phenomena such as misinformed

---

[140] See, e.g., (Gould 2006), (Washington 2007), (Saini 2020), and (Cleghorn 2021).

[141] See, e.g., (Douglas 2009) and (Brown 2024).

[142] See, e.g., (Brown 2004) and (Angell 2005).

non-belief are primarily the result of a dysfunctional epistemic infrastructure and the primary focus should not be on trying to persuade individuals to believe specific scientific claims but on trying to improve our epistemic infrastructure by repairing, upgrading, or redesigning its components. While I cannot further develop this view here,[143] I hope that this Element can nevertheless contribute to this shift in focus by exposing the inadequacies of the science denial framework and more generally of the post-truth diagnosis.

---

[143] See (Contessa 2023) for a sketch of the social approach to public trust in science. For similar approaches to this general topic, see, among many others, (Wilholt 2013), (Levy 2017), (de Melo-Martín and Intemann 2018), (Irzık and Kurtulmuş 2019), (de Cruz 2020), (Goldenberg 2021), (Irzık and Kurtulmuş 2021), (Levy 2021), (Levy 2022), and (Furman 2024).

References

Achen, Christopher H., and Larry M. Bartels. 2016. *Democracy for Realists: Why Elections Do Not Produce Responsive Government*. Princeton, NJ: Princeton University Press.

Almassi, Ben. 2012. 'Climate Change and the Ethics of Individual Emissions: A Response to Sinnott-Armstrong'. *Perspectives: International Postgraduate Journal of Philosophy* 4 (1): 4–21.

Altanian, Melanie. 2024. *The Epistemic Injustice of Genocide Denialism*. London: Routledge.

Alter, Charlotte. 2014. 'Anti-Vaccination Parents Reject Scientific Evidence against Autism'. *Time*, 4 March. https://healthland.time.com/2014/03/04/nothing-not-even-hard-facts-can-make-anti-vaxxers-change-their-minds/.

Anderson, Elizabeth. 2011. 'Democracy, Public Policy, and Lay Assessments of Scientific Testimony'. *Episteme* 8 (2): 144–64. https://doi.org/10.3366/epi.2011.0013.

Angell, Marcia. 2005. *The Truth about the Drug Companies: How They Deceive Us and What to Do about It*. New York: Random House.

Aytaç, Uğur. 2021. 'On the Limits of the Political: The Problem of Overly Permissive Pluralism in Mouffe's Agonism'. *Constellations* 28 (3): 417–31. https://doi.org/10.1111/1467-8675.12525.

Ball, James. 2018. *Post-Truth: How Bullshit Conquered the World*. London: Biteback.

Bardon, Adrian. 2019. *The Truth about Denial: Bias and Self-Deception in Science, Politics, and Religion*. New York: Oxford University Press.

Beres, Derek, Matthew Remski, and Julian Walker. 2023. *Conspirituality: How New Age Conspiracy Theories Became a Public Health Threat*. London: Penguin.

Berman, Jonathan M. 2020. *Anti-Vaxxers: How to Challenge a Misinformed Movement*. Cambridge, MA: MIT Press.

Blewett, Lynn A., Gestur Davidson, Matthew D. Bramlett, Holly Rodin, and Mark L. Messonnier. 2008. 'The Impact of Gaps in Health Insurance Coverage on Immunization Status for Young Children'. *Health Services Research* 43 (5 Pt 1): 1619–36. https://doi.org/10.1111/j.1475-6773.2008.00864.x.

Bogart, Laura M., Glenn Wagner, Frank H. Galvan, and Denedria Banks. 2010. 'Conspiracy Beliefs about HIV Are Related to Antiretroviral Treatment Nonadherence among African American Men with HIV'. *Journal of*

Acquired Immune Deficiency Syndromes (1999) 53 (5): 648–55. https://doi.org/10.1097/QAI.0b013e3181c57dbc.

BonJour, Laurence. 1978. 'Can Empirical Knowledge Have a Foundation?' *American Philosophical Quarterly* 15 (1): 1–14.

Brennan, Jason. 2017. *Against Democracy*. Princeton, NJ: Princeton University Press.

Brown, James Robert. 2004. 'Money, Method and Medical Research'. *Episteme* 1 (1): 49–59. https://doi.org/10.3366/epi.2004.1.1.49.

Brown, Matthew J. 2024. 'For Values in Science: Assessing Recent Arguments for the Ideal of Value-Free Science'. *Synthese* 204 (4): 112. https://doi.org/10.1007/s11229-024-04762-1.

Carter, J. Adam. 2024. 'Epistemic Normativity Is Not Independent of Our Goals'. In *Contemporary Debates in Epistemology, 3rd Ed.*, edited by Blake Roeber, Ernest Sosa, Matthias Steup, and John Turri, 263–73. Hoboken, NJ: Wiley-Blackwell.

Cassam, Quassim. 2019. *Vices of the Mind: From the Intellectual to the Political*. Oxford: Oxford University Press.

Cleghorn, Elinor. 2021. *Unwell Women: Misdiagnosis and Myth in a Man-Made World*. New York, NY: Dutton.

Collins, Harry M., and Robert Evans. 2007. *Rethinking Expertise*. Chicago, IL: University of Chicago Press.

Conee, Earl, and Richard Feldman. 2004. *Evidentialism: Essays in Epistemology*. Oxford: Oxford University Press.

Contessa, Gabriele. 2022. 'Shopping for Experts'. *Synthese* 200 (3): 1–21. https://doi.org/10.1007/s11229-022-03590-5.

Contessa, Gabriele. 2023. 'It Takes a Village to Trust Science: Towards a (Thoroughly) Social Approach to Public Trust in Science'. *Erkenntnis* 88 (7): 2941–66. https://doi.org/10.1007/s10670-021-00485-8.

Public Trust in Science: A Communitarian Approach. Unpublished Manuscript.

Cook, John, Dana Nuccitelli, Sarah A. Green et al. 2013. 'Quantifying the Consensus on Anthropogenic Global Warming in the Scientific Literature'. *Environmental Research Letters* 8 (2): 024024. https://doi.org/10.1088/1748-9326/8/2/024024.

Cook, John, Naomi Oreskes, Peter T. Doran et al. 2016. 'Consensus on Consensus: A Synthesis of Consensus Estimates on Human-Caused Global Warming'. *Environmental Research Letters* 11 (4): 048002. https://doi.org/10.1088/1748-9326/11/4/048002.

Croce, Michel. 2018. 'Expert-Oriented Abilities vs. Novice-Oriented Abilities: An Alternative Account of Epistemic Authority'. *Episteme* 15 (4): 476–98. https://doi.org/10.1017/epi.2017.16.

Cruz, Helen de. 2020. 'Believing to Belong: Addressing the Novice-Expert Problem in Polarized Scientific Communication'. *Social Epistemology* 34 (5): 440–52. https://doi.org/10.1080/02691728.2020.1739778.

D'Ancona, Matthew. 2018. *Post-Truth: The New War on Truth and How to Fight Back*. London: Ebury Press.

Desai, Angel N., and Maimuna S. Majumder. 2020. 'What Is Herd Immunity?' *JAMA* 324 (20): 2113. https://doi.org/10.1001/jama.2020.20895.

Ditto, Peter H., Brittany S. Liu, Cory J. Clark et al. 2019. 'At Least Bias Is Bipartisan: A Meta-Analytic Comparison of Partisan Bias in Liberals and Conservatives'. *Perspectives on Psychological Science* 14 (2): 273–91. https://doi.org/10.1177/1745691617746796.

Donald J. Trump [@realDonaldTrump]. 2012. 'The Concept of Global Warming Was Created by and for the Chinese in Order to Make U.S. Manufacturing Non-Competitive'. Tweet. *Twitter*. https://twitter.com/realDonaldTrump/status/265895292191248385.

Douglas, Heather. 2009. *Science, Policy, and the Value-Free Ideal*. Pittsburgh, PA: University of Pittsburgh Press.

⸺ 2023. 'Differentiating Scientific Inquiry and Politics'. *Philosophy* 98 (2): 123–46. https://doi.org/10.1017/S0031819122000432.

Druckman, James N., and Mary C. McGrath. 2019. 'The Evidence for Motivated Reasoning in Climate Change Preference Formation'. *Nature Climate Change* 9: 111–19.

Ellis, Jon. 2022. 'Motivated Reasoning and the Ethics of Belief'. *Philosophy Compass* 17 (6): 1–11. https://doi.org/10.1111/phc3.12828.

Fagan, Moira, and Christine Huang. 2019. 'A Look at How People around the World View Climate Change'. *Pew Research Center* (blog). 18 April 2019. www.pewresearch.org/short-reads/2019/04/18/a-look-at-how-people-around-the-world-view-climate-change/.

Fantl, Jeremy, and Matthew McGrath. 2009. *Knowledge in an Uncertain World*. Oxford: Oxford University Press.

Fasce, Angelo, Philipp Schmid, Dawn L. Holford et al. 2023. 'A Taxonomy of Anti-Vaccination Arguments from a Systematic Literature Review and Text Modelling'. *Nature Human Behaviour* 7 (9): 1462–80. https://doi.org/10.1038/s41562-023-01644-3.

Feemster, Kristen A., and Claire Szipszky. 2020. 'Resurgence of Measles in the United States: How Did We Get Here?' *Current Opinion in Pediatrics* 32 (1): 139–44. https://doi.org/10.1097/MOP.0000000000000845.

Fine, Cordelia. 2011. *Delusions of Gender: How Our Minds Society and Neurosexism Create Difference*. Reprint edition. New York: W. W. Norton.

Frances, Bryan, and Jonathan Matheson. 2024. 'Disagreement'. In *The Stanford Encyclopedia of Philosophy*, edited by Edward N. Zalta and Uri Nodelman, Winter. Metaphysics Research Lab, Stanford University. https://plato.stanford.edu/archives/win2024/entries/disagreement/.

Frankfurt, Harry G. 2005. *On Bullshit*. Princeton, NJ: Princeton University Press.

Friedman, Lisa. 2020. 'Covid, Climate and Denial'. *The New York Times*, 7 October, sec. Climate. www.nytimes.com/2020/10/07/climate/covid-climate-and-denial.html.

Furman, Katherine. 2024. 'Beliefs, Values and Emotions: An Interactive Approach to Distrust in Science'. *Philosophical Psychology* 37 (1): 240–57. https://doi.org/10.1080/09515089.2023.2266454.

Gaffney, Adam, David U. Himmelstein, Samuel Dickman, Danny McCormick, and Stephanie Woolhandler. 2023. 'Uptake and Equity in Influenza Vaccination among Veterans with VA Coverage, Veterans without VA Coverage, and Non-Veterans in the USA, 2019-2020'. *Journal of General Internal Medicine* 38 (5): 1152–59. https://doi.org/10.1007/s11606-022-07797-7.

Gaffney, Adam W., Steffie Woolhandler, and David U. Himmelstein. 2022. 'Association of Uninsurance and VA Coverage with the Uptake and Equity of COVID-19 Vaccination: January–March 2021'. *Journal of General Internal Medicine* 37 (4): 1008–11. https://doi.org/10.1007/s11606-021-07332-0.

Gerken, Mikkel. 2022. *Scientific Testimony: Its Roles in Science and Society*. Oxford: Oxford University Press.

Goldberg, Sanford C. 2010. *Relying on Others: An Essay in Epistemology*. Oxford: Oxford University Press.

2018. *To the Best of Our Knowledge: Social Expectations and Epistemic Normativity*. Oxford: Oxford University Press.

Goldberg, Jeffrey. 2020. 'Why Obama Fears for Our Democracy'. *The Atlantic*, 16 November. www.theatlantic.com/ideas/archive/2020/11/why-obama-fears-for-our-democracy/617087/.

Goldenberg, Maya J. 2021. *Vaccine Hesitancy: Public Trust, Expertise, and the War on Science*. Pittsburgh, PA: University of Pittsburgh Press.

Gould, Stephen Jay. 2006. *The Mismeasure of Man*. Revised and Expanded edition. New York: W. W. Norton.

Grizzard, Matthew, Rebecca Frazer, and Charles Monge. 2023. 'Demystifying Schadenfreude: How Disposition Theorizing Explains Responses to Social Media Stories of Unvaccinated COVID-19 Deaths'. *New Media & Society* 27: 702–25, July, 14614448231184868. https://doi.org/10.1177/14614448231184868.

Habgood-Coote, Joshua. 2019. 'Stop Talking about Fake News!' *Inquiry: An Interdisciplinary Journal of Philosophy* 62 (9–10): 1033–65. https://doi.org/10.1080/0020174x.2018.1508363.

Hannon, Michael. 2023. 'The Politics of Post-Truth'. *Critical Review: A Journal of Politics and Society* 35 (1): 40–62. https://doi.org/10.1080/08913811.2023.2194109.

Hansson, Sven Ove. 2017. 'Science Denial as a Form of Pseudoscience'. *Studies in History and Philosophy of Science Part A* 63 (June): 39–47. https://doi.org/10.1016/j.shpsa.2017.05.002.

Hares, Andrew, Janet Dickinson, and Keith Wilkes. 2010. 'Climate Change and the Air Travel Decisions of UK Tourists'. *Journal of Transport Geography*, Tourism and climate change, 18 (3): 466–73. https://doi.org/10.1016/j.jtrangeo.2009.06.018.

Hiller, Avram. 2011. 'Climate Change and Individual Responsibility'. *The Monist* 94 (3): 349–68. https://doi.org/10.5840/monist201194318.

Holford, Dawn L., Angelo Fasce, Thomas H. Costello, and Stephan Lewandowsky. 2023. 'Psychological Profiles of Anti-Vaccination Argument Endorsement'. *Scientific Reports* 13 (1): 1–12. https://doi.org/10.1038/s41598-023-30883-7.

Holford, Dawn L., Philipp Schmid, Angelo Fasce, and Stephan Lewandowsky. 2024. 'The Empathetic Refutational Interview to Tackle Vaccine Misconceptions: Four Randomized Experiments'. *Health Psychology* 43 (6): 426–37. https://doi.org/10.1037/hea0001354.

Holpuch, Amanda. 2014. 'Obama Calls for Climate Change Action in California Commencement Speech'. *The Guardian*, 14 June, sec. Environment. www.theguardian.com/environment/2014/jun/14/obama-university-california-irvine-speech-climate-change.

Hoofnagle, Mark, and Chris Jay Hoofnagle. 2007. 'What Is Denialism?' *Denialism* (blog). 30 April 2007. https://denialism.com/about/.

Imhoff, Roland, and Pia Karoline Lamberty. 2017. 'Too Special to Be Duped: Need for Uniqueness Motivates Conspiracy Beliefs'. *European Journal of Social Psychology* 47 (6): 724–34. https://doi.org/10.1002/ejsp.2265.

Irzık, Gürol, and Faik Kurtulmuş. 2019. 'What Is Epistemic Public Trust in Science?' *The British Journal for the Philosophy of Science* 70 (4): 1145–66. https://doi.org/10.1093/bjps/axy007.

——— 2021. 'Well-Ordered Science and Public Trust in Science'. *Synthese* 198 (Suppl 19): 4731–48. https://doi.org/10.1007/s11229-018-02022-7.

Iyengar, Shanto, and Sean J. Westwood. 2015. 'Fear and Loathing across Party Lines: New Evidence on Group Polarization'. *American Journal of Political Science* 59 (3): 690–707.

Jenke, Libby. 2024. 'Affective Polarization and Misinformation Belief'. *Political Behavior* 46 (2): 825–84. https://doi.org/10.1007/s11109-022-09851-w.

Jylhä, Kirsti M., Samantha K. Stanley, Maria Ojala, and Edward J. R. Clarke. 2023. 'Science Denial'. *European Psychologist* 28 (3): 151–61. https://doi.org/10.1027/1016-9040/a000487.

Kahan, Dan M. 2015. 'Climate-Science Communication and the Measurement Problem'. *Political Psychology* 36 (S1): 1–43. https://doi.org/10.1111/pops.12244.

——— 2016. 'The Politically Motivated Reasoning Paradigm, Part 1: What Politically Motivated Reasoning Is and How to Measure It'. In *Emerging Trends in the Social and Behavioral Sciences: An Interdisciplinary, Searchable, and Linkable Resource*, edited by Robert A. Scott and Stephen Kosslyn, 1–16. Hoboken, NJ: Wiley. https://doi.org/10.1002/9781118900772.etrds0417.

——— 2017. 'The "Gateway Belief" Illusion: Reanalyzing the Results of a Scientific-Consensus Messaging Study'. *Journal of Science Communication* 16 (5): 1–20. https://doi.org/10.22323/2.16050203.

Kahan, Dan M., Ellen Peters, Maggie Wittlin et al. 2012. 'The Polarizing Impact of Science Literacy and Numeracy on Perceived Climate Change Risks'. *Nature Climate Change* 2 (10): 732–35. https://doi.org/10.1038/nclimate1547.

Kahn-Harris, Keith. 2018. 'Denialism: What Drives People to Reject the Truth'. *The Guardian*, 3 August, sec. News. www.theguardian.com/news/2018/aug/03/denialism-what-drives-people-to-reject-the-truth.

Kalichman, Seth C., Lisa Eaton, and Chauncey Cherry. 2010. '"There Is No Proof That HIV Causes AIDS": AIDS Denialism Beliefs among People Living with HIV/AIDS'. *Journal of Behavioral Medicine* 33 (6): 432–40. https://doi.org/10.1007/s10865-010-9275-7.

Kourany, Janet, and Martin Carrier, eds. 2020. *Science and the Production of Ignorance: When the Quest for Knowledge Is Thwarted*. Cambridge, MA: The MIT Press.

Kreil, Agnes S. 2021. 'Does Flying Less Harm Academic Work? Arguments and Assumptions about Reducing Air Travel in Academia'. *Travel Behaviour and Society* 25 (October): 52–61. https://doi.org/10.1016/j.tbs.2021.04.011.

Kuhn, Thomas S. 1957. *The Copernican Revolution: Planetary Astronomy in the Development of Western Thought*. Cambridge, MA: Harvard University Press.

Kwon, Diana. 2019. 'How to Debate a Science Denier'. *Scientific American*. www.scientificamerican.com/article/how-to-debate-a-science-denier/.

Lackey, Jennifer. 2018. 'Experts and Peer Disagreement'. In *Knowledge, Belief, and God: New Insights in Religious Epistemology*, edited by Matthew A. Benton, John Hawthorne, and Dani Rabinowitz, 228–45. Oxford: Oxford University Press.

Levin, Dan. 2021. 'They Died from Covid: Then the Online Attacks Started'. *The New York Times*, 27 November, sec. Style. www.nytimes.com/2021/11/27/style/anti-vaccine-deaths-social-media.html.

Levy, Neil. 2017. 'Due Deference to Denialism: Explaining Ordinary People's Rejection of Established Scientific Findings'. *Synthese*, June. https://doi.org/10.1007/s11229-017-1477-x.

——— 2021. 'Echoes of Covid Misinformation'. *Philosophical Psychology* 36 (5): 931–48. https://doi.org/10.1080/09515089.2021.2009452.

——— 2022. *Bad Beliefs: Why They Happen to Good People*. New York: Oxford University Press.

Levy, Neil, and Russell Varley. forthcoming. 'Mind the Guardrails: Epistemic Trespassing and Apt Deference'. *Social Epistemology*. https://doi.org/10.1080/02691728.2024.2400560.

Lewandowsky, Stephan. 2021. 'Liberty and the Pursuit of Science Denial'. *Current Opinion in Behavioral Sciences*, Human Response to Climate Change: From Neurons to Collective Action, 42 (December): 65–69. https://doi.org/10.1016/j.cobeha.2021.02.024.

Lewandowsky, Stephan, John Cook, Ullrich K. H. Ecker et al. 2020. *The Debunking Handbook 2020*. Fairfax, VA: George Mason University Center for Climate Change Communication. https://sks.to/db2020.

Li, Wenchao, Christopher Long, Tongyao Fan et al. 2023. 'Gas Cooking and Respiratory Outcomes in Children: A Systematic Review'. *Global Epidemiology* 5 (December): 100107. https://doi.org/10.1016/j.gloepi.2023.100107.

Lin, Weiwei, Bert Brunekreef, and Ulrike Gehring. 2013. 'Meta-Analysis of the Effects of Indoor Nitrogen Dioxide and Gas Cooking on Asthma and Wheeze in Children'. *International Journal of Epidemiology* 42 (6): 1724–37. https://doi.org/10.1093/ije/dyt150.

Linden, Sander L. van der, Anthony A. Leiserowitz, Geoffrey D. Feinberg, and Edward W. Maibach. 2015. 'The Scientific Consensus on Climate Change as a Gateway Belief: Experimental Evidence'. *PLOS ONE* 10 (2): e0118489. https://doi.org/10.1371/journal.pone.0118489.

Linden, Sander L. van der, Chris E. Clarke, and Edward W. Maibach. 2015. 'Highlighting Consensus among Medical Scientists Increases Public Support for Vaccines: Evidence from a Randomized Experiment'. *BMC Public*

Health 15 (1207): 1–5 (December). https://doi.org/10.1186/s12889-015-2541-4.

Litman, Leib, Zohn Rosen, Rachel Hartman et al. 2023. 'Did People Really Drink Bleach to Prevent COVID-19? A Guide for Protecting Survey Data against Problematic Respondents'. *PLOS ONE* 18 (7): e0287837. https://doi.org/10.1371/journal.pone.0287837.

Mann, Michael, and Tom Toles. 2016. *The Madhouse Effect: How Climate Change Denial Is Threatening Our Planet, Destroying Our Politics, and Driving Us Crazy*. Illustrated edition. New York: Columbia University Press.

Marlon, Jennifer R., Emily Goddard, Peter Howe et al. 2023. 'Yale Climate Opinion Maps 2023'. *Yale Program on Climate Change Communication* (blog). 13 December. https://climatecommunication.yale.edu/visualizations-data/ycom-us/.

Marshall, Barry J., and J. Robin Warren. 1984. 'Unidentified Curved Bacilli in the Stomach of Patients with Gastritis and Peptic Ulceration'. *The Lancet* 323 (8390): 1311–15. https://doi.org/10.1016/S0140-6736(84)91816-6.

McIntyre, Lee. 2018. *Post-Truth*. Cambridge, MA: The MIT Press.

2019. *The Scientific Attitude: Defending Science from Denial, Fraud, and Pseudoscience*. Cambridge, MA: The MIT Press.

2021. *How to Talk to a Science Denier: Conversations with Flat Earthers, Climate Deniers, and Others Who Defy Reason*. Cambridge, MA: The MIT Press.

McKenna, Robin. 2020. 'Persuasion and Epistemic Paternalism'. In *Epistemic Paternalism: Conceptions, Justifications and Implications*, edited by Guy Axtell and Amiel Bernal, 91–106. Lanham, MD: Rowman & Littlefield International.

Melo-Martín, Immaculada de, and Kristen Intemann. 2018. *The Fight against Doubt: How to Bridge the Gap Between Scientists and the Public*. Oxford: Oxford University Press.

Michaels, David, ed. 2008. *Doubt Is Their Product: How Industry's Assault on Science Threatens Your Health*. Oxford: Oxford University Press.

Miller, Boaz. 2013. 'When Is Consensus Knowledge Based? Distinguishing Shared Knowledge from Mere Agreement'. *Synthese* 190 (7): 1293–1316. https://doi.org/10.1007/s11229-012-0225-5.

Mnookin, Seth. 2012. *The Panic Virus: The True Story Behind the Vaccine-Autism Controversy*. New York: Simon & Schuster.

Mooney, Chris. 2014. 'Study: You Can't Change an Anti-Vaxxer's Mind'. *Mother Jones* (blog). 3 March. www.motherjones.com/environment/2014/03/vaccine-denial-psychology-backfire-effect/.

2021. 'Why Science Denial Is Driven by Much More than Just Politics'. *Washington Post*, 27 October. www.washingtonpost.com/news/energy-environment/wp/2015/07/01/politics-may-interfere-with-science-but-not-all-science-denial-is-political/.

Mouffe, Chantal. 2005. *On the Political*. London: Routledge.

Nadler, Steven M., and Lawrence Shapiro. 2021. *When Bad Thinking Happens to Good People: How Philosophy Can Save Us from Ourselves*. Princeton, NJ: Princeton University Press.

Nguyen, C. Thi. 2020. 'Echo Chambers and Epistemic Bubbles'. *Episteme* 17: 141–61. https://doi.org/10.1017/epi.2018.32.

Norman, Andy. 2021. *Mental Immunity: Infectious Ideas, Mind-Parasites, and the Search for a Better Way to Think*. New York: Harper Wave.

NPR. 2015. 'Scientific Evidence Doesn't Support Global Warming, Sen. Ted Cruz Says'. *NPR*. 9 December. www.npr.org/2015/12/09/459026242/scientific-evidence-doesn-t-support-global-warming-sen-ted-cruz-says.

Nyhan, Brendan, and Jason Reifler. 2015. 'Does Correcting Myths about the Flu Vaccine Work? An Experimental Evaluation of the Effects of Corrective Information'. *Vaccine* 33 (3): 459–64. https://doi.org/10.1016/j.vaccine.2014.11.017.

Nyhan, Brendan, Jason Reifler, Sean Richey, and Gary L. Freed. 2014. 'Effective Messages in Vaccine Promotion: A Randomized Trial'. *Pediatrics* 133 (4): e835–e842, February. https://doi.org/10.1542/peds.2013-2365.

Offit, Paul A. 2015. *Deadly Choices: How the Anti-Vaccine Movement Threatens Us All*. Illustrated edition. New York: Basic Books.

Oreskes, Naomi, and Erik M. Conway. 2010. *Merchants of Doubt: How a Handful of Scientists Obscured the Truth on Issues from Tobacco Smoke to Global Warming*. New York, NY: Bloomsbury Press.

Paul, Annie Murphy. 2017. 'Why Science Denial Isn't Necessarily Ideological'. *Washington Post*, 25 May, sec. Opinions. www.washingtonpost.com/opinions/why-science-denial-isnt-necessarily-ideological/2017/05/25/c8cc8346-3f14-11e7-8c25-44d09ff5a4a8_story.html.

Pomerantsev, Peter. 2020. *This Is Not Propaganda: Adventures in the War against Reality*. Illustrated edition. New York: PublicAffairs.

Porter, Amy, and Johanna Goldfarb. 2019. 'Measles: A Dangerous Vaccine-Preventable Disease Returns'. *Cleveland Clinic Journal of Medicine* 86 (6): 393–98. https://doi.org/10.3949/ccjm.86a.19065.

Potochnik, Angela. 2024. *Science and the Public*. Elements in the Philosophy of Science. Cambridge: Cambridge University Press.

Rascoe, Ayesha. 2020. 'Herman Cain, Former GOP Presidential Candidate, Dies From COVID-19'. *NPR*, 30 July, sec. Politics. www.npr.org/2020/07/30/897158910/herman-cain-former-presidential-candidate-dies-from-covid-19.

Raz, Joseph. 1985. 'Authority and Justification'. *Philosophy & Public Affairs* 14 (1): 3–29.

———. 1986. *The Morality of Freedom*. Oxford: Clarendon Press.

Ridder, Jeroen de. 2024. 'What's So Bad about Misinformation?' *Inquiry: An Interdisciplinary Journal of Philosophy* 67 (9): 2956–78. https://doi.org/10.1080/0020174x.2021.2002187.

Rushing, J. Taylor, and David Martosko. 2016. 'Trump Insists It Was a "joke" When He Claimed Global Warning Is a Hoax'. *Daily Mail*, 18 January, sec. News. www.dailymail.co.uk/news/article-3405315/Trump-insists-joke-claimed-Chinese-invented-global-warming.html.

Saini, Angela. 2017. *Inferior: How Science Got Women Wrong – and the New Research That's Rewriting the Story*. Boston, MA: Beacon Press.

Saini, Angela. 2020. *Superior: The Return of Race Science*. Boston, MA: Beacon Press.

Schmid, Philipp, and Cornelia Betsch. 2019. 'Effective Strategies for Rebutting Science Denialism in Public Discussions'. *Nature Human Behaviour* 3 (9): 931–39. https://doi.org/10.1038/s41562-019-0632-4.

Schulz, Kathryn. 2010. 'Stress Doesn't Cause Ulcers! or, How to Win a Nobel Prize in One Easy Lesson: Barry Marshall on Being ... Right'. *Slate*, 10 September. https://slate.com/news-and-politics/2010/09/stress-doesn-t-cause-ulcers-or-how-to-win-a-nobel-prize-in-one-easy-lesson-barry-marshall-on-being-right.html.

Sinatra, Gale, and Barbara Hofer. 2021. *Science Denial: Why It Happens and What to Do about It*. New York: Oxford University Press.

Sinnott-Armstrong, Walter. 2005. 'It's Not My Fault: Global Warming and Individual Moral Obligations'. In *Perspectives on Climate Change*, edited by Walter Sinnott-Armstrong and Richard Howarth, 221–53. Amsterdam: Elsevier.

Smith, Allan. 2021a. 'A Trio of Conservative Radio Hosts Died of Covid: Will Their Deaths Change Vaccine Resistance?' *NBC News*. 3 September. www.nbcnews.com/politics/politics-news/trio-conservative-radio-hosts-died-covid-will-their-deaths-change-n1278258.

Smith, David Livingstone. 2021b. *Making Monsters: The Uncanny Power of Dehumanization*. Cambridge, MA: Harvard University Press.

Specter, Michael. 2009. *Denialism: How Irrational Thinking Hinders Scientific Progress, Harms the Planet, and Threatens Our Lives*. 1st ed. New York: Penguin Press.

Stegenga, Jacob. 2018. *Medical Nihilism*. Oxford: Oxford University Press.

Stott, Peter. 2021. *Hot Air: The Inside Story of the Battle against Climate Change Denial*. London: Atlantic Books.

Tappin, Ben M., Gordon Pennycook, and David G. Rand. 2020a. 'Bayesian or Biased? Analytic Thinking and Political Belief Updating.' *Cognition* 204: 1–12. 10.1016/j.cognition.2020.104375

——— 2020b. 'Thinking Clearly about Causal Inferences of Politically Motivated Reasoning: Why Paradigmatic Study Designs Often Undermine Causal Inference'. *Current Opinion in Behavioral Sciences* 34: 81–87.

Taylor, Kyla W., Sorina E. Eftim, Christopher A. Sibrizzi et al. 2025. 'Fluoride Exposure and Children's IQ Scores: A Systematic Review and Meta-Analysis'. *JAMA Pediatrics* 179 (3): 282–92. https://doi.org/10.1001/jamapediatrics.2024.5542.

Thagard, Paul. 1999. *How Scientists Explain Disease*. Princeton, NJ: Princeton University Press.

Thaller, Annina, Anna Schreuer, and Alfred Posch. 2021. 'Flying High in Academia – Willingness of University Staff to Perform Low-Carbon Behavior Change in Business Travel'. *Frontiers in Sustainability* 2: 1–10, (December). https://doi.org/10.3389/frsus.2021.790807.

Till, Christine, and Rivka Green. 2021. 'Controversy: The Evolving Science of Fluoride: When New Evidence Doesn't Conform with Existing Beliefs'. *Pediatric Research* 90 (5): 1093–95. https://doi.org/10.1038/s41390-020-0973-8.

Trivers, Robert. 2011. *The Folly of Fools: The Logic of Deceit and Self-Deception in Human Life*. New York: Basic Books.

Tsai, George. 2014. 'Rational Persuasion as Paternalism'. *Philosophy and Public Affairs* 42 (1): 78–112. https://doi.org/10.1111/papa.12026.

Tseng, Sherry H. Y., Craig Lee, and James Higham. 2022. 'Managing Academic Air Travel Emissions: Towards System-Wide Practice Change'. *Transportation Research Part D: Transport and Environment* 113 (December): 103504. https://doi.org/10.1016/j.trd.2022.103504.

Tufekci, Zeynep. 2025. 'We Were Badly Misled about the Event that Changed Our Lives'. *The New York Times*, 16 March, sec. Opinion. www.nytimes.com/2025/03/16/opinion/covid-pandemic-lab-leak.html.

Tyson, Alec, Cary Funk, and Brian Kennedy. 2023. 'What the Data Says about Americans' Views of Climate Change'. *Pew Research Center*. www.pewresearch.org/short-reads/2023/08/09/what-the-data-says-about-americans-views-of-climate-change/.

Vickers, Peter. 2023. *Identifying Future-Proof Science*. Oxford: Oxford University Press.

Washington, Harriet A. 2007. *Medical Apartheid: The Dark History of Medical Experimentation on Black Americans from Colonial Times to the Present*. New York: Doubleday.

Watson, Jamie Carlin. 2021. *Expertise: A Philosophical Introduction*. London: Bloomsbury.

West, Mick. 2018. *Escaping the Rabbit Hole: How to Debunk Conspiracy Theories Using Facts, Logic, and Respect*. New York: Skyhorse.

Whitmarsh, Lorraine, Stuart Capstick, Isabelle Moore, Jana Köhler, and Corinne Le Quéré. 2020. 'Use of Aviation by Climate Change Researchers: Structural Influences, Personal Attitudes, and Information Provision'. *Global Environmental Change* 65 (November): 102184. https://doi.org/10.1016/j.gloenvcha.2020.102184.

Wilholt, Torsten. 2013. 'Epistemic Trust in Science'. *British Journal for the Philosophy of Science* 64 (2): 233–53. https://doi.org/10.1093/bjps/axs007.

Williams, Daniel. 2023a. 'The Case for Partisan Motivated Reasoning'. *Synthese* 202 (3): 1–27. https://doi.org/10.1007/s11229-023-04223-1.

———. 2023b. 'The Marketplace of Rationalizations'. *Economics and Philosophy* 39 (1): 99–123. https://doi.org/10.1017/s0266267121000389.

Williams, Joshua T. B., John Rice, Matt Cox-Martin, Elizabeth A. Bayliss, and Sean T. O'Leary. 2019. 'Religious Vaccine Exemptions in Kindergartners: 2011–2018'. *Pediatrics* 144 (6): e20192710. https://doi.org/10.1542/peds.2019-2710.

Worsnip, Alex. 2024. 'Epistemic Normativity Is Independent of Our Goals'. In *Contemporary Debates in Epistemology, 3rd Ed.*, edited by Blake Roeber, Ernest Sosa, Matthias Steup, and John Turri, 253–62. Hoboken, NJ: Wiley-Blackwell.

Wynne, Brian. 1996. 'May the Sheep Safely Graze? A Reflexive View of the Expert-Lay Knowledge Divide'. In *Risk, Environment and Modernity: Towards a New Ecology*, edited by Scott Lash, Bronislaw Szerszynski, and Brian Wynne, 27–83. London: SAGE.

Zagzebski, Linda Trinkaus. 2012. *Epistemic Authority: A Theory of Trust, Authority, and Autonomy in Belief*. Oxford: Oxford University Press.

Acknowledgements

I am very thankful to Heather Douglas, Kevin Elliott, Neil Levy, Ted Richards, and an anonymous reviewer for Cambridge University Press for their insightful and constructive feedback on previous drafts. I would also like to thank Jacob Stegenga for his patience and support throughout this project as well as Melanie Altanian, TY Branch, Matthew J. Brown, Michel Croce, Mikkel Gerken, Joyce Havstad, and Joe Roussos for helpful discussions of these topics. This Element draws on research supported by the Social Sciences and Humanities Research Council of Canada through an Insight Development Grant.

Cambridge Elements

Philosophy of Science

Jacob Stegenga
NTU Singapore

Jacob Stegenga is a Professor at NTU Singapore, and previously taught at the University of Cambridge. He has published widely in philosophy of science and philosophy of medicine, and is the author of *Medical Nihilism*, described as 'a landmark work', *Care and Cure: An Introduction to Philosophy of Medicine*, and a book to be published in 2025 titled *Heart of Science*.

About the Series

This series of Elements in Philosophy of Science provides an extensive overview of the themes, topics and debates which constitute the philosophy of science. Distinguished specialists provide an up-to-date summary of the results of current research on their topics, as well as offering their own take on those topics and drawing original conclusions.

Cambridge Elements

Philosophy of Science

Elements in the Series

Logical Empiricism as Scientific Philosophy
Alan W. Richardson

Scientific Models and Decision Making
Eric Winsberg and Stephanie Harvard

Science and the Public
Angela Potochnik

Feminist Philosophy of Science
Anke Bueter

Abductive Reasoning in Science
Finnur Dellsén

Climate Science
Wendy S. Parker

The Social Dimensions of Scientific Knowledge: Consensus, Controversy, and Coproduction
Boaz Miller

Scientific Realism
Timothy D. Lyons

Science, Pseudoscience, and the Demarcation Problem
Dániel Bárdos and Adam Tamas Tuboly

Underdetermination and Theoretical Virtues
Dana Tulodziecki

The Philosophy of Linguistics
Ryan M. Nefdt

Science Denial: Post-Truth or Post-Trust?
Gabriele Contessa

A full series listing is available at: www.cambridge.org/EPSC

Printed by Integrated Books International,
United States of America